T0206301

Biomolecular
EPR
Spectroscopy

Biomolecular
EPR
Spectroscopy

Wilfred Raymond Hagen

CRC Press
Taylor & Francis Group
Boca Raton London New York

CRC Press is an imprint of the
Taylor & Francis Group, an **informa** business

CRC Press
Taylor & Francis Group
6000 Broken Sound Parkway NW, Suite 300
Boca Raton, FL 33487-2742

First issued in paperback 2020

© 2009 by Taylor & Francis Group, LLC
CRC Press is an imprint of Taylor & Francis Group, an Informa business

No claim to original U.S. Government works

ISBN-13: 978-0-367-57740-7 (pbk)
ISBN-13: 978-1-4200-5957-1 (hbk)

Visit the Taylor & Francis Web site at
http://www.taylorandfrancis.com

and the CRC Press Web site at
http://www.crcpress.com

Dedication

*This book is dedicated to my longtime teachers
S. P .J. Albracht and W. R. Dunham*

Table of Contents

Part 2   Theory

Part 3   Specific Experiments

Preface

Molecular EPR spectroscopy is a method to look at the structure and reactivity of molecules; likewise, biomolecular EPR spectroscopy—*bioEPR* for short—is a method to look at the structure and function of biomolecules. Like every spectroscopy this one also has its specific advantages and limitations. For example, compared to its closest congener, NMR spectroscopy, the applicability of EPR is obviously limited to paramagnetic substances. Therefore, when used in the study of metalloproteins, for example, not the whole molecule is observed, as is the case with proton NMR, but only that small part where the paramagnetism is located. On the other hand, this is usually the central place of action, that is, the active site of enzymic catalysis. Again compared to NMR, EPR then, with its increased concentration sensitivity, becomes a remarkable tool for a focused look into molecular biological action, at least for the vast group of biological transition ion complexes.

The seed for this book was planted many years ago when as an enthusiastic but not particularly focused undergraduate student in chemistry I entered the laboratory of biochemistry of the University of Amsterdam to seek advice on subjects of putative interest for a research project. In a remote corner of the building, two floors below ground level—well below the bordering canal's water level—I entered a cramped room filled with electronic equipment, with control panels high up in the air obviously configured to be operated by preference by the tall man standing in front. When Siem Albracht explained to me in a few words the startling potential of these toys to get directly to the molecular heart of biological activity, I was immediately won over, and therefore only slightly put back by his advice not to return before having spent some time in the physical chemistry department across the canal to learn the basics of magnetic resonance. Over there, I was assigned a desk in an empty room and given a book filled with hundreds of quantum mechanical equations. I reappeared from the room a month later only to receive other books and research papers and lectures equally filled with equations, and the better part of a year later I exited phys chem with a head full of matrices, but with a mind trying to remember why I got interested in this subject in the first place. Fortunately, the enchantment quickly resurfaced after I was finally admitted to biochemistry, and I began an oscillatory journey between the cold room, where biomass (then, bovine hearts) was converted into pure enzymes, and the EPR room, where these enzymes were converted into spectra that challenged one's interpretational skills.

Over the subsequent three decades I have often wondered whether there would not be an alternative road to enter the bioEPR field for those of us, like myself, who choose to work in an intrinsically multidisciplinary area such as biochemistry (or microbiology, coordination chemistry, medical chemistry, et cetera) where some practical and theoretical knowledge is required on a broad range of advanced methods and instrumentation. Could one envision a way to short-cut the impractically time-consuming requirement to work one's way through the physics of EPR without seriously compromising one's final level of expertise? In other words, if not a totally free lunch, could one come up with a quick, budgetary lunch of sufficient nutritional value? A unique chance to not only toss this idea around, but to actually experiment with it at length, offered itself when in the second half of the 1980s Bob Crichton and Cees Veeger started to run their yearly Advanced Course on Metals in Biology at the University of Louvain-la-Neuve, where "bio" and "physics" graduate students from all over Europe were brought together for a 10-day crash course on the methods of bioinorganic chemistry (and on the doubly, triply, or even quadruply fermented beers of Belgium). The EPR spectroscopic experience of most of these students would be limited, at best, to having been permitted to a look over the shoulder of their supervisor at the spectrometer console, and so it became my challenging task to turn them into active bioEPR spectroscopists by means of a 3-hour lecture plus a day hands-on at the spectrometer. The course ran for almost two decades, allowing me to try out and improve my alternative road to EPR enlightenment on close to 700 involuntary guinea pigs.

Obviously, the course would not provide more than a starting point, and those who found themselves really touched by the EPR virus, were expected to continue and dive deeper into the matter using their own strength. At the end of the course many would ask for a book title to further develop their knowledge, and my somewhat embarrassed answer would always be that, although well over a hundred books have been written on the subject, the vast majority of them starts QM and ends QM and has QM in between, and reference to biological systems and to the specific problems of bioEPR is only made occasionally, if at all. And thus this book has been written to fill a void, not to ignore or deny the relevance of quantum mechanics for bioEPR, but to develop a biocentric approach to the problem, in which the experiment, including the biological EPR sample preparation, is the starting point, the spectral interpretation is valued from a point of view of biological relevance, and selected topics from quantum mechanics and its associated matrix algebra may eventually prove to be indispensable, but also relatively easy to deal with for the uninitiated, where the need for their application arises naturally from the practice of bioEPR. In brief, this is a modern version of the book that I would have wanted to have read as an eager student to get a head start in the field.

Wilfred R. Hagen

Department of Biotechnology
Delft University of Technology
Delft, The Netherlands

Part 1

Basics

1 Introduction

1.1 OVERVIEW OF BIOMOLECULAR EPR SPECTROSCOPY

Electron paramagnetic resonance (EPR) spectroscopy, also less frequently called electron spin resonance (ESR) spectroscopy, or occasionally electron magnetic resonance (EMR) spectroscopy, is the resonance spectroscopy of molecular systems with unpaired electrons. Although there are more molecules without unpaired electrons (diamagnets) than with unpaired electrons (paramagnets), the latter are usually of particular interest. For example, biomolecules with unpaired electrons are transition ion complexes or radicals, and these structures are frequently found where the biological action is in the active center of enzymes, for example. This book is about the EPR spectroscopy of biomolecules and of several classes of biochemically relevant synthetic molecules, notably models or mimics, probes, and traps. Biomolecular EPR spectroscopy, or bio-EPR for short, has a long and imposing history as a tool in the life sciences. The technique has been instrumental in the discovery of biological metal clusters, an area of research that in its turn has greatly stimulated the still expanding field of synthetic metal cluster chemistry. Also, bio-EPR has been a key technique in the initial characterization of copper, nickel, molybdenum sites, to name a few, as the hearts of metalloenzymes. And studying radical biochemistry is not easily envisioned without the resource of an EPR machine.

One of the truly fascinating aspects of biomolecular EPR spectroscopy is its interdisciplinary position at the crossroads of biology, chemistry, and physics. The history of bio-EPR tells a story of numerous examples of what, at first sight perhaps, may not have been obvious scientific liaisons but eventually led to scientific discoveries of importance. An early illustrative anecdote is that of the $g \approx 1.94$ EPR line, which today is generally considered to be an almost infallible flag for the presence of iron–sulfur clusters. When the biologists initially suggested that this signal must be related to the presence of ferric ion, the physicists were quick to chastise them for their ignorant revolt against quantum mechanics, which dictates that the g-value of ferric ion, due to the "quenching of orbital angular momentum," can only have a "third-order correction" to the free electron value and, therefore, should certainly not deviate more than 0.01 from g = 2.00. Of course, once they were on speaking terms again, they beautifully made up by combining the biologist's suggestion that ferrous ion and perhaps also nonprotein sulfur might be involved, with the physicist's notion that pairs of metal ions can bind through exchange of valence electrons thereby producing new magnetic properties, leading to their jointly supported model of the biological iron–sulfur cluster. And immediately the chemists were there to top off things by synthesizing from simple chemicals equivalent clusters of the right atom stoichiometry and with comparable magnetism. And subsequently all this spurred an avalanche of research activities continuing to this day with ramifications into a rainbow of disciplines such as human medicine (control of iron homeostasis through clusters), bioorganic stereochemistry (prochirality effectuated through clusters), agriculture (regulation of nitrogen fixation through cluster formation), putative future computer hardware (molecular magnets), and many other fields.

This brief anecdote should serve to illustrate that its extensively interdisciplinary character is not only a strength of bio-EPR but also its Achilles' heel. When the production of significant results requires comparable input efforts from different disciplines, there is an increased chance for the occurrence of time-wasting misunderstandings and errors. A less anecdotic example is the claim—frequently found in physics texts—that sensitivity of an EPR spectrometer increases with increasing microwave frequency. Although this statement may in fact be true for very specific boundary conditions—for example, when "sensitivity" stands for absolute sensitivity of low-loss samples of very small dimensions—when applied in the EPR of biological systems it can easily lead to considerable loss of time and money and to frustration on the part of the life science researcher, because it is simply not true at all for (frozen) solutions of biomolecules.

This book on bioEPR intends to avoid such misunderstandings from the start. The primary goal of a bioEPR spectroscopist is to contribute to an understanding of life at the molecular level. The theory and practice of the spectroscopy is selectively developed as a means to this goal.

1.2 HOW TO USE THIS BOOK AND ASSOCIATED SOFTWARE

We begin with the assumption that you have a background in some part of the life sciences or related fields, and that your familiarity with quantum mechanics and the related mathematics (together abbreviated as QM) may be limited or even nonexistent. It is possible to apply biomolecular EPR spectroscopy in your field of research ignoring the QM part, however, for a full appreciation of the method and to develop skills for its all-round applicability, the QM has to be mastered too.

To allow you a choice of what level of sophistication you want to reach for, the book has been divided in three parts: Part 1 (Chapters 1–6), Basics; Part 2 (Chapters 7–9), Theory; and Part 3 (Chapters 10–14), Selected Topics. The first part does not require any previous knowledge in QM; the math is straightforward, and expressions that come from QM are simply given without derivation. Mastering Part 1 will make you a good operator (at least on paper) and a spectroscopist with limitations in bioEPR. In Part 2 we develop the QM required for bioEPR from scratch, which means that you should be able to read your way through this part even without previous QM experience. If you decide that happiness does not (or does not yet) require knowing about matrix algebra and spin operators, then you can skip Part 2, *except* for the preamble in Section 7.1, which is a summary "for dummies" of Chapters 7–9. The final part, 3, can then be read at different levels of appreciation. Some subjects, for example, spin traps and spin labels, are treated with relatively sporadic allusions to QM, and if you just boldly jump over these, you can still get to the straightforward expressions used for most practical problems. Furthermore, learning by way of the human mind is rather different from filling up a linear memory array, and there is nothing wrong with jumping back and forth, that is, acquiring knowledge in a patchy way, and later (even much later) trying to fill in the holes.

This book comes with a suite of programs for basic manipulation and analysis of EPR data, such as constructing frequency-normalized difference spectra, spin counting by integration, simulation of a variety of powder spectra, and rhombogram analysis. All programs are freely available and downloadable from www.bt.tudelft.nl/ biomolecularEPRspectroscopy. Description of the programs and instructions for their use are also to be found there, and not in this book, to avoid outdating and to allow for repeated updates and extensions. The programs have been set up with a view to ease of use: the graphical user interface typically consists of a single window and is designed to be self-explanatory as much as possible.

All programs use input and output files (experimental and simulated spectra) consisting of 1024 amplitude values in a single column in ASCII format. If you have experimental files in a different format, then you must first modify them. A program is included to change from n to 1024 points.

All programs are intended to be run as application on a PC using the Windows operating system (XP or later). The code was written in FORTRAN 90/95 and compiled with the INTEL Visual FORTRAN compiler integrated in the Microsoft Visual Studio Developer Environment.

1.3 A BRIEF HISTORY OF BIOEPR

History—in my view—is an *interpretation* of the past in terms of directional events culminating in the present. This definition implies history is colored by an evaluation of the present. Here is my evaluation of the bioEPR present: contemporary biomolecular EPR spectroscopy is heavily dominated by experiments in X-band (i.e., a microwave frequency of circa 9–10 GHz) on randomly oriented dilute biomolecules in (frozen) aqueous solutions. Perhaps the first and foremost goal of the game is the quantitative identification and monitoring of functional molecular substructures, such as the active site in a metalloprotein, in a manner not conceptually dissimilar to the application of optical spectroscopy to chromophoric biomolecules. This evaluation makes bioEPR quite distinct from (or, if you wish, complementary to) biomolecular crystallography or structural biomolecular Nuclear Magnetic Resonance (NMR) spectroscopy, for example. From this vantage point I read the history of bioEPR in the following chronological quantum leaps.

The electron paramagnetic resonance effect was discovered in 1944 by E. K. Zavoisky in Kazan, in the Tartar republic of the then-USSR, as an outcome of what we would nowadays call a purely curiosity-driven research program apparently not directly related to WW-II associated technological developments (Kochelaev and Yablokov 1995). However, a surplus of radar components following the end of the war did boost the development of EPR spectroscopy, in particular, after the X-band ("X" meaning to be kept a secret from the enemy) was entered in Oxford, U.K., in 1947 (Bagguley and Griffith 1947).

Application to biomolecules started as early as the mid-fifties with single-crystal EPR studies on hemoglobin (Bennett et al. 1955), but in hindsight it now appears that this pioneering work has not led to the most successful development in the evolution of bioEPR. Consider, for example, the following quotation from a second paper (Bennett and Ingram 1956):

> The investigations of single crystals of hemoglobin derivatives by paramagnetic resonance can give two distinct types of information. First, the actual resonance conditions and the resultant g values associated with electronic transitions will yield *details on the orbitals involved in the chemical binding* of the central iron atom. Secondly, the angular variation of the g values enables an accurate determination to be made of the orientation of the haem and porphyrin planes with respect to the external crystalline axes. Although the structure of the rest of the molecule cannot be analyzed directly in this way, detailed information on the orientation of the haem plane can be combined with x-ray measurements to *calculate the polypeptide chain directions and similar factors*. It would appear that the determination of the haem plane orientations by paramagnetic resonance is *much more accurate than that by any other method* so far applied.

The italics are mine. They here expressed the hope that a program in bioEPR would predominantly afford (1) detailed electronic information and (2) detailed 3-D structural data. This expectation is still frequently held and voiced up to this day, however, more than half a century of bioEPR history points to the success of rather more down to earth applications to frozen solution samples for purposes such as metal identification, determination of oxidation state, and stoichiometry of centers in complex systems (for example, respiratory chains). This more biochemically oriented branch of bioEPR traces back to the work of Sands on Fe^{III} centers in glasses (which later turned out to have spectra quite similar to those from frozen solutions of iron proteins such as hemoglobin and rubredoxin), in which an analytical equation is developed to describe the angular variation of the g-value in samples of randomly oriented molecules, which in turn provides the basis for quantitative analysis (e.g., by spectral simulation) of so-called powder patterns (Sands 1955). A crucial period was in the late fifties when biochemist Helmut Beinert in Madison, Wisconsin, regularly took the train to Ann Arbor, Michigan (yes, there was one!), to take his samples of bovine heart cells and mitochondria to physicist Dick Sands to discover a Cu^{II} signal in cytochrome oxidase (Sands and Beinert 1959) and the g = 1.94 signal (Beinert and Sands 1960) of what only many years later was identified as the signature of iron–sulfur clusters. In the same period Bo Malmström, Tore Vänngård, and collaborators in Göteborg, Sweden, started their EPR experiments on frozen solutions of purified metalloproteins (Malmström et al. 1959, Bray et al. 1959).

An obvious requirement for quantitative bioEPR (i.e., determining concentrations from integrated spectra) would be a proper expression for the intensity, or transition probability, and its variation in randomly oriented samples. For a long time the expressions developed by physicists (Bleaney 1960, Kneubuhl and Natterer 1961, Holuj 1966, Isomoto et al. 1970) did not quite seem to work for spectra with significant g-anisotropy, until the matter was finally settled by Roland Aasa and Tore Vänngård in what may well be the most cited paper in bioEPR (Aasa and Vänngård 1975). They made the simple but crucial point that previous expressions were derived for frequency-swept spectra, while the overwhelming practice had become to record field-swept spectra, and this required a correction to the intensity equal to $1/g$, which in informal settings is usually referred to as "the Aasa factor."

For a balanced historical record I should add that the late W. E. Blumberg has been cited to state (W. R. Dunham, personal communication) that "One does not need the Aasa factor if one does not make the Aasa mistake," by which Bill meant to say that if one simulates powder spectra with proper energy matrix diagonalization (as he apparently did in the late 1960s in the Bell Telephone Laboratories in Murray Hill, New Jersey), instead of with an analytical expression from perturbation theory, then the correction factor does not apply. What this all means I hope to make clear later in the course of this book.

In regular EPR experiments the microwave propagation vector is perpendicular to the magnetic field vector. The study of integer spin systems additionally requires a setup (and theory) in which these two vectors are parallel. Parallel-mode EPR was originally introduced for single crystals doped with $J =$ even lanthanide ions in Oxford (Bleaney and Scovil 1952), and it was later applied to randomly oriented organic triplet ($S = 1$) radicals in the Royal Shell laboratory in Amsterdam (Van der Waals and De Groot 1959). In 1982 in Amsterdam I introduced the method to bioEPR in a study of frozen solutions of $S = 2$ metalloproteins and models (Hagen 1982b).

The above historical outline refers mainly to the EPR of transition ions. Key events in the development of radical bioEPR were the synthesis and binding to biomolecules of stable spin labels in 1965 in Stanford (e.g., Griffith and McConnell 1966) and the discovery of spin traps in the second half of the 1960s by the groups of M. Iwamura and N. Inamoto in Tokyo; A. Mackor et al. in Amsterdam; and E. G. Janzen and B. J. Blackburn in Athens, Georgia (e.g., Janzen 1971), and their subsequent application in biological systems by J. R. Harbour and J. R. Bolton in London, Ontario (Harbour and Bolton 1975).

The development of a wide range of special forms of EPR was initiated when the idea of double resonance (using simultaneous irradiation by two different sources) was cast in 1956 by G. Feher at Bell Telephone Labs in his seminal paper on ENDOR, electron nuclear double resonance (Feher 1956). BioEPR applications of ENDOR were later developed on flavoprotein radicals in a collaboration of A. Ehrenberg and L. E. G. Eriksson in Stockholm, Sweden, and J. S. Hyde at Varian in Palo Alto, California (Ehrenberg et al. 1968), and on metalloproteins in a joint effort of the groups of R. H. Sands in Ann Arbor, I. C. Gunsalus in Urbana, Illinois, and H. Beinert in Madison (Fritz et al. 1971).

Perhaps the most noteworthy of this brief historical outline is that all the cited dates are from more than a quarter century ago. Of course, this is not to imply that nothing has happened since in terms of theoretical or technological developments, but the message is that EPR in general, and bioEPR in particular, is a mature spectroscopy, whose application readily pays off if you just take the trouble of getting acquainted with its now-well-defined requirements, possibilities, and limitations.

2 The Spectrometer

This chapter is a guided tour of the standard EPR spectrometer. The goal is not to give a rigorous description of the underlying physics, but to develop a feel for basic parts and principles sufficient to make you an independent, intelligent operator of any X-band machine.

2.1 THE CONCEPT OF MAGNETIC RESONANCE

Spectroscopy requires a source of radiation, a sample, and a detector; magnetic spectroscopy additionally requires an external magnetic field. The term *spectroscopy* implies that at least one of these four elements is variable, or tunable, in some way or other, and that one measures the amount of radiation absorbed by the sample as a function of this variable. For example, the source generates radiation with energy

$$E = h\nu \qquad (2.1)$$

in which h is Planck's constant, and ν is the frequency (in units of Hertz or cycles per second) with its corresponding wavelength λ (in meters) according to the conversion

$$\lambda = c / \nu \qquad (2.2)$$

in which c is the speed of light (2.99792×10^8 meters per second). Tuning the frequency over a limited range of the electromagnetic spectrum is the most common approach taken in the majority of spectroscopies. Dealing with different ranges of the EM spectrum requires different technologies, and therefore each range has its own spectroscopy (or spectroscopies), from very low-frequency (i.e., very low energy and very long wavelengths) radio waves in NMR spectroscopy to very high-frequency (i.e., very high energy and very short wavelengths) gamma rays in x-ray spectroscopy. In magnetic spectroscopy, one has the alternative possibility to vary the magnetic field while keeping the frequency at a constant value. This is the approach usually taken in EPR spectroscopy. On the contrary, in other magnetic spectroscopies—for example, NMR, MCD (magnetic circular dichroism), and MS (Mössbauer spectroscopy)—the magnetic field is kept constant and the frequency is varied. In principle, one can, of course, also vary both the field and the frequency at the same time, but this is rarely done. The choice of what to vary is always based on practical considerations of technical limitations, which we will discuss for EPR later.

Figure 2.1 illustrates the concept of magnetic resonance in EPR spectroscopy. The sample is a system that can exist in two different states with energies that are degenerate (i.e., identical) in the absence of a magnetic field but that are different in the presence of a field—for example, a molecule with a single unpaired electron.

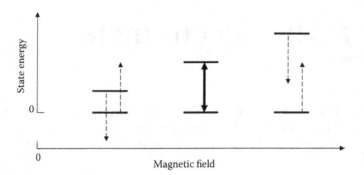

FIGURE 2.1 The concept of magnetic resonance. A degenerate two-level system is split in a magnetic field. The energy difference between the two states increases with increasing field, and this affords its tuning to fit the energy of an electromagnetic wave of fixed frequency (the vertical bar). The leftmost level scheme is below resonance, the middle scheme is at resonance, and the rightmost scheme is above resonance.

The difference in energy is a function of the strength of the external magnetic field. The term "external" means that the field is not produced by the sample itself, but by an external device, such as an electromagnet. The strength of the field is used to tune the energy difference of the two molecular states such that it becomes exactly equal to the energy of incoming radiation from the source. The radiation can now be used for the transition of molecules from one state to the other, that is, from the lower to the higher state (absorption) and from the higher to the lower state (stimulated emission; Göppert-Mayer 1931). The term *resonance* refers to this going back and forth between the two states. Normally, more molecules are in the lower state (ground state) than in the higher state (excited state), and resonance will therefore result in net absorption of radiation. This holds for all forms of spectroscopy, however, when the energy difference between two states is large, there may be negligibly few molecules in the excited state, and the term *resonance* is not used, as for example in optical spectroscopy. In EPR spectroscopy the energy difference is about four orders-of-magnitude less than in visible-light spectroscopy, and the populations of the two levels are comparable—hence, electron paramagnetic *resonance*.

The resonance condition for a two-level system in EPR is

$$h\nu = g\beta B \tag{2.3}$$

in which the energy of the radiation produced by the source is equated to the energy difference between the two molecular states produced by the external magnet. The interaction between the compound and the magnetic field is called the *electronic Zeeman interaction*. The equation contains two constants, h and   (i.e., quantities with invariant values given by nature), two variables, ν and B (i.e., two quantities whose values can be chosen by the experimenter, for example, by turning knobs on a spectrometer), and a proportionality constant, g, whose value is the result of the experiment (i.e., the carrier of physical–chemical information). In bold terms, the goal of an EPR experiment is the determination and chemical interpretation of the value of g (and of related quantities for more than two-level systems to be discussed later). Planck's constant h is a universal constant with value 6.62607×10^{-34} J $\times$ s (joule second); the Bohr magneton is a derived electromagnetic constant with value 9.27401×10^{-24} J $\times$ T^{-1} (joule per tesla). The frequency of radiation, v, is in Hz (hertz or cycles per second; units: per second); the external magnetic field (also called *magnetic flux density*), B, is in tesla. The observable g is dimensionless, which is easily seen by rearrangement of Equation 2.3

$$g = \frac{hv}{\beta B}\left[\frac{Jss^{-1}}{JT^{-1}T}\right] \tag{2.4}$$

EPR spectrometers use radiation in the giga-hertz range (GHz is 10^9 Hz), and the most common type of spectrometer operates with radiation in the X-band of microwaves (i.e., a frequency of circa 9–10 GHz). For a resonance frequency of 9.500 GHz (9500 MHz), and a g-value of 2.00232, the resonance field is 0.338987 tesla. The value $g_e = 2.00232$ is a theoretical one calculated for a free unpaired electron in vacuo. Although this esoteric entity may perhaps not strike us as being of high (bio) chemical relevance, it *is* in fact the reference system of EPR spectroscopy, and thus of comparable importance as the chemical-shift position of the ^{1}H line of tetramethylsilane in NMR spectroscopy, or the reduction potential of the normal hydrogen electrode in electrochemistry.

A derived SI-unit for magnetic field is the gauss, which is defined in tesla units as

$$10,000[G] \equiv 1[T] \tag{2.5}$$

The human mind appears to have a preference for employing units that force the value of things into small numbers (i.e., of the order of the number of fingers on two human hands). Therefore, EPR spectroscopists prefer the gauss over the tesla; an EPR linewidth of, say, 8 gauss somehow sounds easier to deal with than a linewidth of 0.0008 tesla. Alternatively, some prefer a linewidth of 0.8 mT (milli-tesla). In this vein, Equation 2.4 is frequently written in the practical form

$$g = 0.714484 \frac{v[MHz]}{B[G]} \tag{2.6}$$

and a free electron in vacuo subject to radiation with a frequency of 9500 MHz resonates at a field of 3389.87 G. What is the energy of this radiation (and therefore the energy difference between the two levels)? Planck's constant was defined in joules-second, so the energy at 9500 MHz is

$$hv = 6.626 \times 10^{-34}\left[Js\right] \times 9500 \times 10^6[s^{-1}] = 6.295 \times 10^{-24}\left[J\right] \tag{2.7}$$

a very small number, indeed. In order to once more satisfy our preference to deal with values close to unity, EPR spectroscopists commonly write energies in units of "reciprocal centimeters" or cm^{-1} using the conversion

$$1[J] \leftrightarrow 5.0348 \times 10^{22}[cm^{-1}] \tag{2.8}$$

which for a frequency of 9500 MHz then gives

$$h\nu = 0.31694[cm^{-1}] \tag{2.9}$$

As an added advantage of using this unit one immediately sees what the wavelength of the employed radiation is

$$\lambda = 1 / 0.31694[cm^{-1}] = 3.1552[cm] \tag{2.10}$$

a number that we will reencounter, below, as $\lambda/2 \approx 16$ mm in the physical dimensions of the spectrometer's microwave components and in the size of the sample!

2.2 THE MICROWAVE FREQUENCY

Why is an X-band microwave of approximately 9.5 GHz a common frequency in EPR spectroscopy? Or, what determines the choice for a specific frequency? To answer this question we look back into the early history of magnetic resonance spectroscopy to the birth and early childhood of NMR and EPR. The very first experiments in both spectroscopies were done in the mid-1940s, rather less by choice than by practical limitations at what we now consider unusually low (not to say impractically low) frequencies. In subsequent developments, ν was steadily increased to get better resolution and higher sensitivity. To understand this requires (1) the concepts of inhomogeneous broadening and (2) the concept of equilibrium populations. We will discuss these concepts in detail later in Chapter 4; for now we simply state their main experimental manifestations: First, EPR absorption lines can have a width that is independent of (or more, generally, *less* than linearly dependent on) the used frequency and the corresponding resonance field. As a consequence, the resolution of two partially overlapping lines will increase with increasing frequency as illustrated in Figure 2.2. Furthermore, an increased resonance field means an increased energy separation between the ground and excited state, which implies an increased surplus of molecules in the ground state available for absorption, and this means an increased EPR amplitude as illustrated in Figure 2.3.

In this game of frequency, pushing to get better resolution and better sensitivity NMR and EPR have marched parallel for a few years, but eventually they have taken highly diverging courses. Up till today, and presumably continuing in the near future, there has been a constant drive in NMR spectroscopy to increase the resonance frequency and field, and this development has only been limited by our technical abilities to construct superconducting magnets of sufficient stability, homogeneity, and strength. On the very contrary, EPR spectroscopists have found an optimum of sensitivity (and, to a lesser extent, of resolution) versus resonance frequency, which is approximately at X-band, say 8–12 GHz (Bagguley and Griffith 1947). The existence of this general optimum has several causes, both of technical and of fundamental nature, and we will consider them in due course. For the time being the bottom line is: all EPR studies start at X-band; increasingly deviating from X-band almost always means an increasing loss of sensitivity, and the added difficulty of experiments outside X-band is only acceptable if extra information is obtainable in addition to what X-band spectroscopy provides (Hagen 1999).

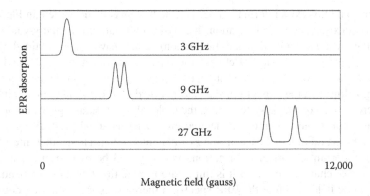

FIGURE 2.2 Resolution may increase with increasing frequency. A two-line EPR absorption spectrum is given at three different microwave frequencies. The line splitting (and also the line position) is caused by an interaction that is linear in the frequency; the linewidth is independent of the frequency. This is a theoretical limit of maximal resolution enhancement by frequency increase. In practical cases the enhancement is usually less; in some cases there is no enhancement at all.

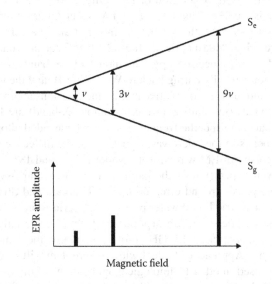

FIGURE 2.3 Sensitivity may increase with increasing frequency. A degenerate two-level system is split in a magnetic field and brought to resonance with three different frequency/field combinations. The EPR amplitude of a single-line spectrum increases (nonlinearly) with increasing frequency as a result of an increased population difference between the states S_g and S_e. This is a highly idealized example of a system with a frequency-independent linewidth and a spectrometer that performs equally well at all three frequencies.

What could this extra information be? For the two-level model system in Figure 2.1 there is *no* extra spectral information: the basic information (i.e., the g-value characterizing the electronic Zeeman interaction) could have been obtained at *any* frequency, and so we obviously choose the most sensitive setup: the X-band spectrometer. The electronic Zeeman interaction is an example of an interaction whose strength is linear in the strength of the external field (and linear in the magnitude of the microwave frequency). Systems (i.e., the molecules of a paramagnetic compound) that can occur in more than two molecular energy levels are always subject to other interactions in addition to the electronic Zeeman interaction. These other interactions contain important chemical information, and they will be treated in considerable depth in later chapters. For now it is important to note that they are *in*dependent of the magnetic field, and that the EPR spectrum therefore is the result of a combination of the Zeeman term linear in B and of one or more terms independent of B. By consequence, changing the microwave frequency means changing the relative weight of the B-dependent and the B-independent interactions, and so the shape (and information content) of the spectrum changes with frequency. For such systems EPR spectroscopy at two or more frequencies (called multifrequency EPR) is worth the extra experimental effort, and is frequently mandatory for unequivocal interpretation. Possible examples are systems with more than one unpaired electron (i.e., high-spin systems) and/or systems with magnetic nuclei (i.e., systems with hyperfine interactions). A typical approach is to start in X-band and then to try a frequency that is a few times less than or greater than X-band. However, recent years have also witnessed exploration of much higher frequencies up to the terahertz range (THz is 10^{12} Hz). That part of the EM spectrum is not considered to be part of the microwaves but rather of the far infrared and the corresponding EPR spectrometers are called "quasi optical." Table 2.1 lists frequency ranges (frequency bands) employed in EPR spectroscopy. The names of the bands originate in military communication partially dating back to World War II, and the nomenclature is unfortunately not unique (i.e., alternative names and divisions are in use). The more commonly used bands (in addition to the ubiquitous X-band) are L-, S-, Q-, and W-bands. Note that some specific frequencies are to be avoided altogether because they encompass peaks in the microwave spectrum of the dielectric (i.e., absorbing the electric component of EM waves) water vapor (22.24 and 183.3 GHz) or in the zero-field microwave spectrum of the paramagnetic molecular oxygen gas (118.7 GHz and multiple peaks around circa 60 GHz) (Tinkham and Strandberg 1955). Furthermore, the dielectric liquid water poses a very serious problem in EPR spectroscopy, because its microwave absorption spectrum is essentially a very broad single band that peaks at circa 2.5 GHz (not accidentally, the same frequency as in microwave ovens). Approaches to deal with this problem will be discussed later. Several bands are used in other technological applications of microwaves, notably, cellular phones (L-band), magnetrons (S-band), airport radars (X-band), satellite televisions (X-band), meteorological radar (S-, C-, K-band), highway speed control (X-, K-band), military communications (several bands), pain rays for crowd control (W-band), and astronomy (the whole spectrum).

TABLE 2.1

Frequency bands in EPR spectroscopy

Band name	Typical frequency ν in GHz	Wavelength λ in mm	Energy hν in reciprocal cm	Resonance field B at g = 2 in tesla
L-band	1	300	0.033	0.036
S-band	3	100	0.10	0.11
C-band	6	50	0.20	0.21
X-band	10	30	0.33	0.36
P-band	15	20	0.50	0.54
K-band	24	12.5	0.80	0.86
Q-band	35	8.6	1.2	1.25
U-band	50	6.0	1.7	1.78
V-band	65	4.6	2.2	2.32
E-band	75	4.0	2.5	2.68
W-band	90	3.3	3.0	3.22
F-band	110	2.7	3.7	3.93
D-band	130	2.3	4.3	4.64
G-band	180	1.67	6.0	6.43
J-band	270	1.11	9.0	9.64
No name	600	0.50	20	21.4
No name	1000	0.30	33	35.7

2.3 OVERVIEW OF THE SPECTROMETER

Figure 2.4 is a schematic drawing of the main components of a standard X-band EPR spectrometer. Let us first walk through this overview, and then look at the components in detail. On the left is a monochromatic source of microwaves of constant output (200 mW) and slightly (10%) tunable frequency, either a klystron or a Gunn diode. The produced radiation is transferred by means of a rectangular, hollow waveguide to an attenuator where the 200 mW can be reduced typically by a factor between 1 and 10^6. The output of the attenuator is transferred with a waveguide to a circulator that forces the waves into the downward waveguide to reach the resonator containing the sample. Just before the wave enters the resonator it encounters the iris, a device to tune the amount of radiation reflected back out of the resonator. The reflected radiation returns to the circulator, where it is forced into the right-hand waveguide to a diode for the detection of microwave intensity. Any remaining radiation that reflects back from the detector is forced by the circulator into the upward waveguide that ends in a wedge, or taper, or "choke" to convert the radiation into heat, so that no radiation can return to the source. A small amount (1% or 20 dB) of the 200 mW source output is "coupled out" before the attenuator to enter the waveguide of the reference arm, which contains a port that can be closed, followed by a device to shift

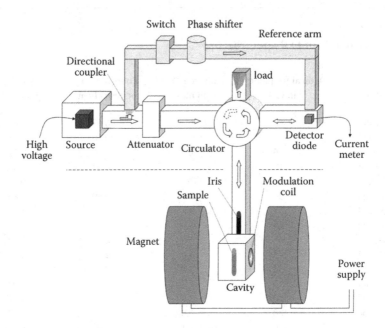

FIGURE 2.4 Schematic drawing of an X-band EPR spectrometer.

the phase of the wave. The output of the reference arm goes directly to the detector diode to produce a constant working current.

Everything above the dotted line is usually not visible because it is built in "the bridge," a rectangular case the size of a boot box. This bridge is placed on a high table approximately at eyesight of a tall, standing person. From a hole in the table, the waveguide with the resonator protrudes downwards. The resonator hangs in the gap of a dipolar electromagnet, between the north and the south pole. The latter are connected to a regulated power supply. The spectrometer produces an xy output (chart and/or file) with the strength of the magnetic field on the x-axis, and the strength of the detector current on the y-axis. Omitted from the figure are the operator's console (the spectrometer's knobs), the water cooling system to stabilize the magnet, the optional "cryogenics" (to cool the sample), and the essential modulation system (a device for the improvement of signal-to-noise, which we will treat separately).

Let us now have a look at the operator interface. The knobs of the spectrometer can be all real physical knobs, or all switches in the software of the spectrometer's computer, or a mixture of both. Real knobs can all be arranged in a separate console, or part of them may have been placed on the front and back panel of the microwave bridge. The computer software can be passive (just collecting data upon a trigger signal) or active (also setting, regulating, and calibrating the spectrometer). Older spectrometers may either have a computer with passive software or no computer at all (only producing a chart on a real xy or xt recorder). Newer spectrometers usually come with active computer control. For rapid orientation on an unfamiliar spectrometer it is useful to look for subcollections of knobs (either real ones or in the software), grouped according to a few basic functions. The functions to set and their main parameters in parentheses are the microwave (frequency, attenuated intensity), the magnet (center field, scan range), the modulation system (strength, frequency), the amplifier (gain), and the data recorder (scan rate and damping time constant). The temperature of the optional cooling (or heating) system is set on a separate control system.

Let us now consider the individual components in some detail (Poole 1996; Czoch and Francik 1989), and then go through the tuning and operating procedure of the spectrometer.

2.4 THE RESONATOR

The heart of the EPR spectrometer is the resonator, also called the cavity, sample holder, or probe head. The term *resonator* is not an allusion to the spectroscopic phenomenon of resonance transitions (cf. Figure 2.1) but to the fact that we use the device in a way comparable to an organ pipe; the EPR resonator is machined according to rather precise dimensional specifications, allowing us to set up a pattern of standing microwaves in its interior. As with an organ pipe, the resonator is used to single out and to amplify a particular frequency. An organ pipe produces a basal tone (its lowest frequency) plus a spectrum of overtones (a specific set of higher frequencies). The equivalence of a musical tone in microwave technology is called a *mode*. EPR resonators are employed as single-mode devices: they amplify only one single "tone," which can be its basal tone or any of its overtones. This mode selection is brought about by injecting monochromatic microwaves into the resonator from an accurately tunable narrow-band source (e.g., an X-band source). The X-band resonator is almost always either a rectangular box or a cylindrical box. At other frequencies one can encounter more esoteric designs (in some very-high-frequency EPR spectrometers, the resonator is abolished altogether).

Figure 2.5 shows the popular rectangular X-band resonator designed to operate in the so-called TE_{102} mode. Microwaves are transported from the source to the resonator as transverse electromagnetic (TEM) waves, which simply means that the sinusoidal variation of the wave is perpendicular to the direction of propagation (in contrast to longitudinal waves that vary along the propagation direction). TE_{102} then means that the wave becomes standing in the resonator with its electrical component having lengths ($\lambda/1$) in the x-direction, undetermined in the y-direction, and ($\lambda/2$) in the z-direction. In an enclosed standing-wave pattern, the magnetic component of the wave wraps itself around the electric-component pattern (and vice versa), with the result that in the middle of the resonator along the x-axis the magnetic component is maximal and the electric component is minimal. This is the ideal position for our bar-shaped sample, because EPR transitions are caused by the magnetic component of the microwave, B_1, which should be perpendicular to the external magnetic field, B, along the z-axis. Any absorption of the electric component of the microwave will be nonresonant (nonspecific) and should be avoided as much as possible because it implies a loss of sensitivity. The choice of the resonator's "undetermined" y-dimension is not completely free; it has a maximum limit determined by the diameter of the magnet poles, and it has a minimum limit determined by the diameter of the cylindrical sample. Also, the strength of the tone of the resonating microwave, also known as the quality factor, Q, of the resonator, is a function of its spatial dimensions

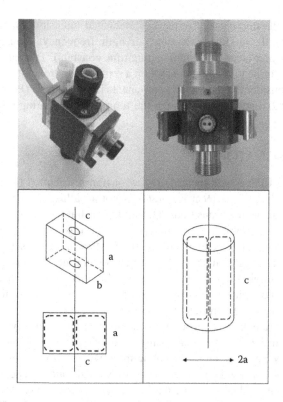

FIGURE 2.5 X-band resonators. A rectangular (left: Bruker 4102ST standard cavity) and a cylindrical (right: Varian E235 large sample access cavity) resonator. The drawings show the critical dimensions for a single-mode standing wave pattern. The broken lines indicate the B_1 field lines in the rectangular TE_{102} mode and in the cylindrical TE_{011} mode.

including the y-dimension. Q also depends on the frequency and on the material of which the resonator is made.

The formal definition of this quality factor, Q, is the amount of power stored in the resonator divided by the amount of power dissipated per cycle (at 9.5 GHz a cycle time is $1/(9.5 \times 10^9) \approx 100$ picoseconds). The dissipation of power is through the resonator walls as heat, in the sample as heat, and as radiation reflected out of the resonator towards the detector. The cycle time is used in the definition because the unit time of one second would be far too long for practical purposes; within one second after the microwave source has been shut off, all stored power has long been dissipated away completely.

Good X-band resonators mounted into a spectrometer and with a sample inside have approximate quality factors of 10^3 or more, which means that they afford an EPR signal-to-noise ratio that is over circa three orders of magnitude better than that of a measurement on the same sample without a resonator, in free space. This is, of course, a tremendous improvement in sensitivity, and it allows us to do EPR on biomolecules in the sub-μM to mM range, but the flip side of the coin is that we are stuck with the specific resonance frequency of the resonator, and so we cannot vary the microwave frequency, and therefore we have to vary the external magnetic field strength.

In modern NMR spectroscopy, the external magnet has a fixed value, and the source of EM radiation is a pulsed one, which means that the sample is irradiated with a whole spectrum of frequencies, and the response of the sample is Fourier-transformed to obtain an absorption spectrum as a function of these frequencies. Why is this approach not taken in EPR spectroscopy? The main reason is that if one would try to take an EPR spectrum at a fixed magnetic field with a pulse of microwave frequencies, one would find a typical spectrum to be circa three orders of magnitude wider in frequency distribution than a typical NMR spectrum. It is technically not very possible to build a microwave source that produces pulses of sufficient intensity and of sufficiently short duration to generate the frequency spectral width that covers a full EPR spectrum. It *is* possible to generate a spectral width that covers a very small part of an EPR spectrum, and this approach is taken in some forms of double resonance spectroscopy, notably pulsed ENDOR (electron-nuclear double resonance) and ESEEM (electron spin echo envelope modulation) to resolve very small splittings from magnetic nuclei. On the contrary, regular EPR spectrometers always use a monochromatic source of continuous waves (CW), in combination with a scanning magnet.

Cylindrical single-mode resonators are also frequently employed in X-band and at higher frequencies (the world record presently is 275 GHz). The most commonly used mode is the one with the lowest frequency: TE_{011}, which also has maximum B_1 along the full y-axis length as illustrated in Figure 2.5. Low overtones are also used, for example, the TE_{012} and TE_{013} modes (one and two nodes in B_1 along the y-axis) in the so-called large access cavity for gas phase X-band EPR. Another example of a low-overtone resonator can be found in the kitchen: microwave ovens are rectangular boxes to store microwaves of 2.45 GHz frequency for dielectric heating of foodstuff. This corresponds (cf. Equation 2.2) to a wavelength of 12.24 cm, and since the diameter of the average pizza does not fit this length, the dimensions of the oven are expanded to multiples of $\lambda/2 \approx 6$ cm, and the mode pattern cannot be the ground mode. As a consequence, the standing wave has nodes at mutual distances of circa 6 cm, and homogeneous heating of the pizza requires a turning table and also a device called a mode stirrer to partially destroy the regularity of the pattern of nodes. For conceptual insight, a microwave oven could be used as a very high-power (1000 W), very low resolution S-band EPR spectrometer; and an X-band EPR spectrometer (≤ 0.2 W) could be viewed as an impractically low-power microwave warmer (to a few degrees above ambient temperature) for mini pizzas of 1 cm diameter. Yet another "household" example is the receiver for satellite television: microwaves emitted by a satellite in a geostationary orbit over the equator are reflected by a disc of 60–100 cm and focused on a device cryptically named "LNB," which stands for low noise amplifier and frequency block converter. The focusing point is a "horn" receiver, a circle that tapers down into a rectangular box looking suspiciously similar to the rectangular X-band resonator of Figure 2.5. It is in fact slightly smaller because the satellite emits in the high end of X-band 11–13 GHz. After reception, the X-band frequencies are as a block down-converted to S-band so that they can be transported through a cheap coaxial cable to the decoder, where the signal is converted to video so that it can be transported through an even cheaper coax cable to the TV set.

2.5 FROM SOURCE TO DETECTOR

What is the source of the microwaves? Monochromatic microwaves are traditionally produced by a vacuum tube called a klystron or more recently by a solid state device known as a Gunn diode. At X-band these devices typically have a maximal output power of 200 mW and an operating life time of the order of 10,000 hours. Usually, sources are "leveled," which means that their power is actually circa 400 mW, but this is "leveled" to exactly 200 mW. When the actual power reduces over the source's life time, its leveled power remains constant until the actual power would eventually drop below the leveled value. Klystrons and Gunn diodes are tunable over a relatively narrow frequency range. Tunability is required because the mode frequency of the resonator becomes *less* when things are placed in its interior (contrary to an organ pipe). For example, an empty X-band resonator with a TE_{102} frequency of 9.8 GHz may resonate at 9.4 GHz after we place a cooling system in it, and at 9.2 GHz when we also add a sample tube. This means that the microwave source should be tunable over an approximate range of 9.1–9.9 GHz. Gunn diodes are produced for frequencies in the range of circa 1–100 GHz. At the price of a significant reduction in power, Gunn diodes can be made to produce multiples of their basic frequency, and this property is used in very high frequency EPR spectrometers. For example, a W-band 90 GHz Gunn diode can be made to produce 180 GHz, 270 GHz, 360 GHz, etc., at ever-decreasing power.

The waves that exit the source must be transported and attenuated. At low frequencies microwaves can be transported through coaxial cables, however, above approximately 6 GHz power losses become unacceptable, and for all practical purposes we have to change to waveguides. These are usually rectangular, but sometimes also cylindrical, empty tubes typically made of brass. The frequency of the radiation to be transported determines the dimensional requirements of the waveguide. For rectangular waveguide a useful approximate rule of thumb is that the broad *external* dimension, A, approaches (is slightly less than) the wavelength λ of the transported wave (cf. Figure 2.6) and the other external dimension is $B \approx A/2$. This is *not* based on any law of physics, because for the spectroscopy, only the *inner* dimensions count, and the outer dimensions depend on the (arbitrary) thickness of the wall, but it does

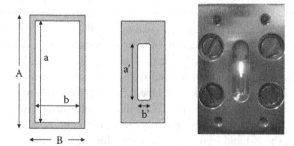

FIGURE 2.6 X-band rectangular waveguide. Dimensions of the waveguide (left); dimensions of the coupling hole (middle); and picture of the coupling hole with the iris screw (right).

happen to come out this way (i.e., $A \approx 0.9\lambda$) for waveguide of practical thickness. In other words, just by looking at an EPR spectrometer from some distance one can easily figure out in what frequency band the machine is operating. The real physics, however, is in the inner dimensions a and b: the length of the long inner side, a, is the cutoff wavelength λ_a of the lowest frequency transportable through the waveguide, the length of the short inner side, b, is the cutoff wavelength λ_b of the highest frequency transportable (in the primary transverse magnetic mode TM_{01}). In practice a rectangular waveguide can be operated in primary mode approximately over the frequency range $1.2\, v_a$–$0.85\, v_b$ (cf. Figure 2.6).

It is perhaps useful to mentally picture the microwaves to travel through the waveguide like a water stream through a pipe. In reality, however, the transport is an electric phenomenon that occurs in a very thin layer of the waveguide's inside. The thickness of this layer is characterized by the skin depth parameter, δ, which depends on the used material and the frequency. For example, for the material copper and a frequency of 10 GHz the skin depth is $\delta \approx 0.66\ \mu m$. While at the surface the amplitude of the electric field of the wave is maximal, at a depth of δ the E is reduced by a factor $e^{-1} \approx 0.37$, and at a depth of a few δ becomes negligibly small. Transmission of microwaves through a waveguide is essentially a surface phenomenon.

The waveguide carries the microwaves to the resonator, which at X-band is a hollow metallic enclosure or cavity. In order for the microwaves to enter into the cavity, one of its end walls must have an opening, which is called the *coupling hole* or *iris*. The "coupling" indicates that this is not just any hole, but that once more dimensions have to be carefully chosen to ensure that all incident power actually gets into the cavity. The common X-band solution is a rectangular hole with extremely thin walls and with dimensions a' and b' that satisfy the conditions $2a' > \lambda$ and $a'/a > b'/b$ as outlined in Figure 2.6. On the waveguide side, just before the iris hole one usually finds a threaded Teflon rod that ends in a small metal plate (see Figure 2.6). Moving this plate up or down by screwing the rod changes the amount of coupling. Only for one unique position the cavity is "critically" coupled, meaning that all power enters the cavity, and no radiation is reflected out (i.e., power is only lost by heat dissipation through the cavity walls). The rod must be tunable because what position of the plate exactly corresponds to critical coupling depends on the cavity and on its contents (sample tube and cryogenics). Frequently, the name "iris" is colloquially (but incorrectly) used for the tunable rod; perhaps the better name "iris screw" should be used. The rod is operated from a distance via an elongation rod either by hand or by means of an electromotor.

On their way from the source to the resonator the intensity of the microwaves must be attenuable for two reasons: (1) full power may be too much for the sample leading to saturation (treated in Chapter 4); or (2) it may be impossible to critically couple the cavity at full incident power (e.g., because the sample contains too much water). Therefore, the main waveguide contains an attenuator, usually of the dissipative, rotary-vane type. *Dissipative* means that the eliminated power is converted to heat and is not reflected as radiation to the source. *Rotary vane* means that it contains a section of circular waveguide, in which a flat piece of material is located that can be rotated over an angle θ, where $\theta = 0$ means no attenuation and $\theta = 90°$ causes full attenuation. The amount of attenuation is expressed in decibels, a non-SI,

TABLE 2.2

Output power conversion from milliwatt to decibel

P (dB)	Pout (mW)
0	200
−1	159
−2	126
−3	100
−4	79.6
−5	63.2
−6	50.2
−7	39.9
−8	31.7
−9	25.2
−10	20.0

logarithmic unit that expresses the ratio of two numbers, or more specifically the magnitude of a physical quantity relative to a given reference level. Here, the reference level is, of course, the full power of the source (e.g., P_{in} = 200 mW). The decibel is a dimensionless unit. Attenuated output power is given by

$$P[dB] = 10 \times \log_{10}\left(P_{out}[mW] / P_{in}[mW]\right) \tag{2.11}$$

For example, when P_{in} = 200 mW is attenuated to P_{out} = 20 mW, then the attenuator indicates an output power P = −10 dB (see also Table 2.2). And when P_{out} = 2 mW, then P = −20 dB. Every additional attenuation by a factor of 10 gives another −10 dB in P. This type of device is usable to 50–60 dB attenuation, which also corresponds to the lower power limit at which the spectrometer is still operable. Note that, in addition to the relative dB scale, microwave engineers also use an absolute dBm scale. The "m" in dBm is shorthand for milliwatt and defines the reference point of this scale:

$$1[dBm] \leftrightarrow 1[mW] \tag{2.12}$$

and so 10 dBm corresponds to 10 mW, 20 dBm means 100 mW, etc. For example, the regulated output power of an X-band spectrometer is usually 200 mW, which on the dBm scale would correspond to $10 \times \log(200)$ dBm. The dBm scale is not commonly used in bioEPR spectroscopy.

2.6 THE MAGNET

The resonance field for a free electron (g_e = 2.0) in X-band is some 3400 gauss (cf. Table 2.1). Higher fields are required for g < 2.0. The vast majority of biological EPR studies will not require detection of g-values less than approximately 1.2

(an explanation will be given in Chapter 5) corresponding to a field of circa 5700 gauss. A versatile X-band EPR spectrometer has a magnet that can be scanned from 0–6000 gauss (0–0.6 tesla). Such fields are readily produced with electromagnets (i.e., two bars of soft iron each surrounded by a coil of copper wire, which are built in a water cooling device). The leads of the coils are attached to a high amperage power supply, which is usually also coupled to the water cooling system, in particular, to dissipate heat produced by a regulatory end stage of power transistors. Also the high-voltage unit that drives the microwave source in the bridge is connected to this cooling system.

With pole diameters of circa 10–20 cm, that is, well above the circumference of the resonator, magnets of this type produce an axial field at the sample in the resonator of sufficient homogeneity for all conceivable EPR purposes. Contrary to the situation in NMR spectroscopy, fine tuning of the field, or shimming, is not required.

Very low fields are not easy to produce. Most electromagnet power supplies will only switch on with a zero offset so that the field starts scanning at a threshold value of 50–100 gauss. In the study of integer spin systems (e.g., high-spin Fe^{II}) scanning through zero field can be of value. This requires an addition to the regulatory power supply: a "through-zero field unit," which will feed a current into the magnet coils of opposite sign to that of the main regulator.

The maximum field attainable with electromagnets is approximately 30,000 gauss, or 3 tesla. This is an absolute limit set by the physics of the soft iron. To reach this maximum requires magnets of considerable dimension and with conal pole caps attached to the poles. In practice, electromagnets are commonly used in EPR spectroscopy up to Q-band (35 GHz). Higher frequencies, in particular from W-band onwards, require superconducting magnets (< 22 T). Extremely high frequencies, above circa 0.5 THz, require special large-scale facilities with resistive magnets or hybrids of superconducting and resistive magnets.

2.7 PHASE-SENSITIVE DETECTION

When an experiment involves the application of a slowly varying entity (slow with respect to the time required to collect a data point), the technique of phase-sensitive detection may apply. In continuous wave EPR spectroscopy the external field is such an entity; it is varied so slowly, compared to the time required to collect an EPR intensity data point, that it is frequently called the "static" field (in contrast to the magnetic component of the microwave that varies with a frequency of 9.5 GHz, for example). If one now applies field modulation, which is a minor disturbance to the static field, that varies in time with an intermediate frequency of, say, 100 kHz, and one passes the signal of the detection diode through an amplifier that "looks" at the signal with exactly the same frequency of 100 kHz (i.e., that looks 10^5 times per second) and with the same phase as the disturbing signal, then the output of the spectrometer differs in two ways from a nondisturbed measurement: (1) the spectrum has a much better signal-to-noise ratio, because all noise that is out of phase with the disturbance is not amplified, and (2) the spectrum is the derivative of the original spectrum (i.e., the derivative of the EPR absorption). Phase-sensitive detection is an absolute requirement for practical biomolecular EPR spectroscopy. On the

spectrometer we must set the amplitude, the frequency, and the phase of the disturbance, or modulation field. How do we choose these quantities?

The modulation is a small, time-varying magnetic field produced by two coils that
are usually built in, or built onto, or placed right next to, the side walls of the resonator
(Figure 2.4). Increase in signal-to-noise is linear in the magnitude of the modulation
field, so one should use the highest possible value. However, there are two limitations
to this value, a spectroscopic and an engineering one. The value of the modulation field
should be significantly (i.e., a few times) less than the linewidth of the measured spectral feature. If the modulation field is comparable to, or even exceeding, the linewidth,
then the signal will be "overmodulated" or deformed by overmodulation. Qualitatively,
this effect may be appreciated as follows: Suppose the modulation field is twice the
linewidth. Then, when the static field is at the center of the line, the modulated field will
some of the time (e.g., at maximal amplitude) be outside the EPR line and no intensity
is detected. The net result is the detection of a broadened line with correspondingly
reduced amplitude as illustrated in Figure 2.7 (note that the total area under the peak *is*
still linear in the applied modulation amplitude). The second limitation comes from the
fact that the coils of the modulating field exert a time-varying force on the side walls
of the resonator, which causes a time-variation of the resonator frequency, and for high
modulation amplitudes this destabilizes the spectrometer. In X-band this destabiliza-
tion defines a practical limit of circa 25 gauss. Typical modulation amplitudes used in
bioEPR are circa 10 gauss for transition ions and circa 1 gauss for radicals.

Increase in signal-to-noise is also linear in the frequency of the modulation. Again,
there is a spectroscopic and an engineering limitation. Let us start with the latter one.
Low-frequency waves (cf., the rumble of an earthquake) penetrate further into matter

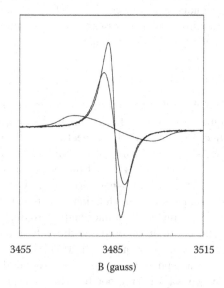

3455   3485   3515

B (gauss)

FIGURE 2.7 Overmodulation. The single-line spectrum of the "strong pitch" calibration
sample ($g = 2.0028$) is recorded at $v = 9.77$ GHz with modulation amplitudes of 2.5, 10, or
40 gauss and with accordingly adjusted electronic gain such that in the absence of modulation
deformation the signal should have constant amplitude.

than high-frequency ones of the same amplitude. This also holds for magnetic-field modulation. A modulation frequency of the order of 100 kHz is approximately the practical upper limit for which the modulation field can still penetrate through the resonator walls (and through an optional cryogenic system) to reach the sample without significant attenuation. Fortunately, this upper engineering limit is associated with a negligible spectroscopic problem. At 9.5 GHz a free electron resonates at $B_r \approx 3400$ gauss. The 100 kHz modulation generates artifacts called spectral "side bands" at field shifts ΔB_r equal to

$$\left| \Delta B_r \right| = \frac{100 \text{kHz}}{9.5 \text{GHz}} \times 3400 \text{gauss} \approx 0.04 \text{gauss} \tag{2.13}$$

which is well within the experimental inhomogeneous EPR linewidth of biomolecules. In summary, 100 kHz is generally the optimal value for the modulation frequency in bioEPR. The only exception is the recording of very easily saturable signals, for example, at very low temperatures, where, in order to avoid a condition known as "rapid passage," the modulation frequency is sometimes lowered by one or two orders of magnitude.

Finally, the phase of the modulating field has to be set, but this is "duck soup": The signal amplitude of an arbitrary sample is experimentally maximized by adjusting the phase. And for a given modulation-coil setup (usually associated with a particular resonator) this has to be done only once in the setup's life time.

Phase-sensitive detection is not at all specific for EPR spectroscopy but is used in many different types of experiments. Some readers may be familiar with the electrochemical technique of differential-pulse voltammetry. Here, the potential over the working and reference electrode, E, is varied slowly enough to be considered as essentially static on a short time scale. The disturbance is a pulse of small potential difference, ΔE, and the in-phase, in-frequency detection of the current affords a very low noise differential of the i-E characteristic of a redox couple.

2.8 TUNING THE SPECTROMETER

It is hands-on-experience time: let us tune the spectrometer for optimal performance, and then take a spectrum. The cooking recipe in Table 2.3 is spelled out for an old fashioned all-hardware X-band spectrometer with 200 mW maximal output power.

TABLE 2.3

Operation instruction sheet

STARTUP (once)

1. Turn on the cooling water (to magnet coils, magnet power supply, microwave bridge).

2. Turn on the main electrical switch(es) (to the magnet, the console).

3. Open the drying gas (filtered pressurized air, or any inert gas) to the waveguide just above the resonator.

(Continued)

TABLE 2.3 (CONTINUED)
Operation instruction sheet

4. Optionally, turn on the regulated temperature device and select a temperature (kelvin).
5. Switch on the controlling computer or (optionally) switch on the "passive" computer.
6. Switch on the microwave bridge to tune mode (and allow two minutes warm-up time).

Tuning (each sample and each temperature change)

7. (Re)place a sample in the spectrometer.
8. Optionally, adjust temperature.
   Set the block of frequency parameters. (Can possibly be skipped for a sequel of samples in identical sample tubes.):
9. Set microwave power to tuning value (typically, −24 dB to get a full scale scope mode).
10. Set the frequency to fit the resonator (center the "dip" in the scope mode; cf. Figure 2.8abc).
11. Tune for critical coupling (zero the "dip" in the scope mode; cf. Figure 2.8d).
12. Set microwave power to reference tuning value (typically −40 dB to get a full scale scope mode).
13. Open the reference arm and set the reference phase (to symmetrical scope mode; cf. Figure 2.8e).
14. Switch the microwave bridge to "operation mode."
15. Set the detector bias current to mid-scale (typically 200 µA).
16. Check coupling (by power variation) and optionally adjust the iris (to power-invariant bias).
17. Choose a microwave power (e.g., −16 dB as a first guess).

Setting (each spectrum)

   Set the field block parameters:
18. Set the center field (e.g., 3400 gauss).
19. Set the field scan (e.g., 1000 gauss for a metal or 100 gauss for a radical).
20. Set the scan time (e.g., 200 s).
21. Set the time constant for noise damping (typically 1/1000th of scan time, e.g., 0.2 s; cf. Figure 2.9).
    Set the signal parameters:
22. Choose a modulation amplitude (at 100 kHz and optimal phase).
23. Choose an amplifier gain (to match signal to paper or screen or AD converter).
    Record a spectrum:
24. Optionally, adjust modulation, gain, scan time, damping.
25. Optionally, optimize power setting (to maximal nonsaturating value; see Chapter 4).

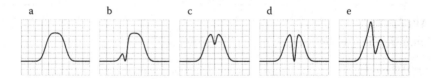

FIGURE 2.8 Tuning mode patterns. The scope mode shows a "dip" when the frequency of the microwave fits into the resonator. This dip has maximal depth when the spectrometer is tuned (with the adjustable iris) to be reflectionless, that is, when the resonator is critically coupled to the bridge. Shown patterns are (a) off resonance, (b) slightly off resonance, (c) either under- or over-coupled, (d) critically coupled, (e) asymmetry from out-of-phase reference.

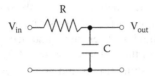

FIGURE 2.9 The low-pass filter. The time constant $\tau = RC$, and the cutoff frequency $\nu_c = 1/(2\pi\tau)$.

For more recently produced instruments some, or many, of the handling steps may actually have to be carried out in silico.

Here are some additional remarks on the startup procedure:

- The drying gas is to prevent condensation of water from air in the resonator, and to exclude air (oxygen) from the resonator.
- Some spectrometers have a controlling computer, that is, setting (and sometimes calibration) of parameters such as field strength is done via switches in the software (soft knobs). Spectrometers that are set with hard knobs on a console may be connected to a "passive" computer (a computer that waits for a trigger signal ["start spectrum recording"] to digitally collect and file data). When a regulating computer is out of order, the spectrometer is "dead"; when a passive computer is out of order, the spectrometer can still output to a chart recorder.
- When the microwave bridge is in tune mode, the microwave source is at high voltage, and its guaranteed lifetime is ticking away (therefore, switch to "off" for a lunch break).
- Regarding the tuning procedure, note the following: When a frozen sample is replaced by a new one from a liquid-nitrogen storage, then the new sample must be wiped dry of condensed air with, for example, a cotton shirt to avoid interfering oxygen signals (see Chapter 3).
- Setting the block of frequency parameters can sometimes be omitted for subsequent "similar" samples if the spectrometer is sufficiently stable. Similar samples means that the sample tubes come from a single batch of quartz tubing (i.e., they have identical inner and outer diameter within a small tolerance) and the samples are physically similar (e.g., frozen buffered protein solutions). Such samples are invariant as dielectric, and they all will cause the same shift in resonator frequency within circa 1 MHz.

2.9 INDICATIVE BUDGET CONSIDERATIONS

What is the price of an EPR machine, and what does it cost to run it? We will quote a few numbers, which should be appreciated as not more than ballpark indicators with obvious time and place dependence. The numbers are in euros for no other reason than the author's location. Multiply by 1.5 for approximate numbers in dollars; divide by 1.5 for approximate numbers in pounds; and multiply by 5/3 to read kiloyen. As a mental exercise, replace "EPR machine" with "automobile" to realize that the answer can be anywhere between a trivially small and an astronomically large number. If your goal is to get from A to B, and appearance is less of an issue, then buying a car can be a modest investment, and driving costs are dominated by gas usage, which are linear in the distance and travel frequency between A and B. However, if you are into big, shiny things and you are of the type to be challenged by the road between A and B being a mud track with no legal speed limit, regular commuting may become a financial matter of concern. A prospective EPR spectroscopist is in a similar situation. A 20-year-old X-band machine may be obtained by no more than the costs of a mover (it does weigh a couple of hundred kilograms), but it may still be a state-of-the-art research instrument due to the remarkable fact that sensitivity in X-band EPR has not significantly (well, OK, a factor of two, perhaps) improved over the last four decades. And the "driving" costs are dominated by usage of cryogenic gasses, that is, linear in the number of low-temperature experiments. On the contrary, a new X-band machine has a price tag of 200 kEuro or more; and setting up, running, and maintaining a 1 THz spectrometer may approach the price range of running a commercial airliner. Let us look at the "low end" in some more detail, and for increased insight let us split total costs into set-up investment (get a spectrometer), bench fees (run the machine), maintenance (fix it when it's broke), and hidden costs (square meters of laboratory space; a dedicated operator).

To evaluate set-up costs we assume that we have to start from scratch. From our previous discussion about microwave frequencies it should be obvious that we want a cw X-band spectrometer as the central (frequently: only) facility. What exactly is a complete spectrometer? The answer depends a bit on the type of experiments planned, but for all cases the minimum requirements would be a basic spectrometer (bridge + resonator; magnet; control unit) *and* a frequency counter.

A magnetic-field meter would be a rather useful, but not absolutely necessary addition. With reference to Equation 2.3 we concluded that there are two parameters to be set experimentally, frequency v and field B, to determine the unknown, g. This implies that we must be able to calibrate both v and B. However, employing a standard sample with a known EPR spectrum would reduce the problem to a need to calibrate either v or B. We go for frequency determination, because this is a quantity that we have to measure constantly as it may change with every experiment in a way that we cannot easily control (it depends on the resonator plus cooling system plus sample tube dimensions plus sample dielectric properties plus modulation amplitude plus temperature). On the other hand, the choice of B settings is ours, and when the magnet is properly working, the values only depend on our settings. Note that required accuracy of v-B pairs depends on the g-value resolution, which depends on the class of compounds studied: radicals (high), $S = 1/2$ metals (not so high), $S > 1/2$ metals (even less high). Also, as noted before: field reproducibility requirements in EPR are not nearly as high as in NMR. A good calibrated electromagnet will always do the job, and no shimming is required.

All experiments except those on spin labels and spin traps require a cryogenic sample-cooling system. The purchase of cryogenic gasses is by far the most significant entry on the budget for running costs. A ballpark number would be 10–20 kEuro annually for the total use of liquid nitrogen and liquid helium of a reasonably active research unit. The cheap and easy solution is the finger dewar to be filled with a small amount

(50 cc) of liquid nitrogen; see Figure 2.10. This provides a constant sample temperature of 77 K with no chance of variation. A temperature of 77 K is too high for rapidly relaxing paramagnets, so this device is only useful for frozen radicals and for some $S = 1/2$ metals. In addition to its inflexibility, it has the disadvantage of producing noisy spectra because bubbles of boiling nitrogen are a jittering source of slight resonator detuning. The latter problem can be rather considerably reduced by bubbling minute amounts of helium from a gas cylinder through narrow tubing into the boiling nitrogen a centimeter or two below the nitrogen surface. A slightly more involved solution is a nitrogen flow system with a boiler and heater as in Figure 2.10. The sample temperature is from circa 85 K up to ambient and above, but for frozen solutions, the upper limit is circa 175 K, above which the sample's dielectric constant is raised sufficiently to destroy the Q factor of the loaded resonator (in simple words, no EPR because all microwaves are absorbed by the water). An advantage of a flow system is a much more stable signal (no bubbles); liquid nitrogen usage is of the order of 10–20 liters per day (1 liter costs circa 1 Euro). Significant bills begin to appear when liquid helium is required (1 liter costs circa 10 Euro), as happen to be the case for the majority of metalloprotein studies. Purchase of a commercial He-flow system may also be a significant set-up cost entry. Over the years we have been charmed by the cheap and simple "Swedish system" named after its country of birth, where it was developed in the Gothenburg lab of Vänngård in the early 1970s (Lundin and Aasa 1973, Albracht 1974). It consists of a home-blown, 85 cm straight quartz dewar that is placed through the resonator right into a 30-liter liquid helium vessel under an elevated magnet (Figure 2.10). Mounting and demounting of this system is a matter of minutes. Note that all helium flow systems require pumping with a combined rotation-diffusion pump on a daily basis before mounting in order to restore proper vacuum.

 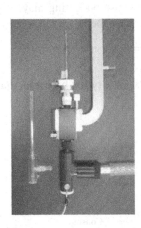 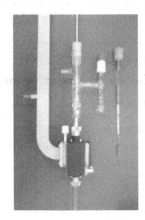

FIGURE 2.10 Sample cooling systems. Left: cold finger for liquid nitrogen. Middle: Bruker N_2 flow system also showing a duplicate of the central dewar; the cold nitrogen gas comes from the right (liquid nitrogen vessel is not shown); a thermocouple comes in from below. Right: home-built He-flow dewar ("Swedish system") also showing a duplicate of the sample tube with elongation rod; the bottom part goes to the liquid helium vessel.

A microwave power meter may perhaps be a handy gadget every now and then to check output levels of microwave sources. On the other hand, we have seen that power leveling is used to retain a constant output power over 10000 operating hours (many years), and if the unleveled power has dropped to below the level specification, then the source can be expected to leave for eternal hunting grounds any time now.

At the risk of embarrassing some of my physics friends, I would also venture that trying to do bioEPR without some wet lab facility very close by (as in next door) is like trying to commute with a car that runs on a rare type of fuel. Even if you can induce a herd of befriended biologists to readily provide you with just about any interesting sample that you can think of, your own next door wet lab cannot be omitted. It should at least contain a gas manifold (and an inert gas cylinder) for anaerobic sample preparation (discussed in the next chapter) plus some basic equipment like microsyringes, pipettes, magnetic stirrer, balance. Another quite useful gadget is a redox titration cell (discussed in Chapter 13). The minimal bioEPR survival kit also contains a few specific chemicals, for example, for reduction and oxidation of samples (see sample preparation in Chapter 3). Overall costs are modest provided, of course, that the real estate of a wet lab is available.

And finally some real good news: extensive and sophisticated bioEPR data analysis can be done with the PC on your desk or lap with the software that comes with this book.

Table 2.4 summarizes the above in terms of a shopping list for new items. It is once again emphasized that the numbers are to be understood as indicative (e.g., not for use in grant proposals). And recall that old spectrometers are not necessarily inferior at all to new ones, and they can be very cheap if standing in someone's way. Also, items such as frequency counters up to X-band appear regularly as second hand offers on the Internet for a fraction of their new price. And finally, the table does not show possible hidden costs, that is, of items that are taken for granted because they already happen to be around, but whose budgeting may be prohibitive when they have to be acquired, for example, square meters of lab space or dedicated operators.

TABLE 2.4
Indicative costs of an X-band EPR facility for biomolecular spectroscopy in k€

Cost type	Class	Item	Price (k€)
Acquisition costs			
	Spectrometer		
		Microwave bridge	50
		Resonator	15
		Magnet	90
		Power supply	15
		Console	20
	Peripherals		
		Hook-up	p.m.[a]
		Frequency counter	15

TABLE 2.4 (CONTINUED)
Indicative costs of an X-band EPR facility for biomolecular spectroscopy in k€

Cost type	Class	Item	Price (k€)
		Field meter	15
		Power meter	8
	Cryogenics		
		Cold finger (nitrogen)	1
		Regulated flow system (nitrogen)	15
		Liquid nitrogen supply vessel	15
		Regulated flow system (helium)	20
		Liquid helium supply vessel	15
		The Swedish system (helium)	1
		Rotation-diffusion pump	4
Lab costs			
		Storage vessel (nitrogen)	4
		Vacuum manifold	1
		Survival kit	3
Running costs			
	Liquid nitrogen (4000 liter per year)		4/y
	Liquid helium (400 liter per year)		4/y
Repair costs			
		Replacement microwave source	15
		Refurbishing resonator walls	7
Hidden costs			
	Lab space		p.m.
	Dedicated personnel		p.m.

Note: 1€↔1.5$; 1€↔(2/3)£; 1€↔(5/3)k¥

[a] *p.m. (pro memoria = just to remind one): costs that cannot be specified because they strongly depend on local conditions.*

3 The Sample

In this chapter we go through some practical aspects on how to prepare an EPR sample. An optimal protein sample for EPR has a volume of circa 250 μL and a protein concentration of the order of 1–100 mg/mL, that is, as high as possible compliant with the protein's solubility and availability. An EPR sample tube is a hollow quartz cylinder of circa 0.5 × 14 cm sealed on one end and with a label at 1 cm from the other end. A reduced-size version with an inner diameter of 1 mm can be used for aqueous samples. Freezing and thawing EPR samples requires strict adherence to specific handling protocols which are different for aqueous samples and nonaqueous samples. Low-temperature bioEPR spectroscopy is nondestructive; thawing the sample affords full recovery of biological activity. Low-temperature spectroscopy on frozen solutions of electron-transfer proteins characterizes a conformation corresponding to the freezing point of the protein solution. The chapter ends with a few important notes on safety.

3.1 SAMPLE TUBE AND SAMPLE SIZE

BioEPR samples are generally (frozen) aqueous solutions since water is the only solvent compatible with terrestrial life. The high-frequency dielectric constant of ice is circa 30 times less than that of water. As a consequence liquid-phase EPR is experimentally rather different from frozen-solution EPR. We start with a discussion of sample handling for low-temperature experiments.

Standard rectangular X-band TE_{102} resonators have a circular access of circa 11 mm diameter (cf. Figure 2.5). In principle a cylindrical sample tube of 10 mm outer diameter could smugly fit into the resonator. In practice, this geometry would not be a good idea at all for two reasons. Firstly, the cryogenic cooling system is usually put inside the resonator, so it requires space at the expense of the sample tube diameter. Secondly, the distribution of the magnetic and the electric component of the microwave for the TE_{102} mode are such that a frozen aqueous sample in a 10 mm tube with, say, an inner diameter of 8–9 mm would experience an electric field of considerable intensity to the extent that all microwaves would disappear in the sample as in a black hole and spectroscopy would be impossible. Figure 3.1 gives a cut-through top view indicating limiting dimensions in practice: the inner tube diameter can not exceed 4 mm. For a sample height of $h = 1.5$ cm this gives a cylindrical volume of circa 200 μL:

$$\text{Sample volume} = \pi \, r^2 \, h = 3.14 \times 0.2 \times 0.2 \times 1.5 \approx 0.2 \text{ mL} \qquad (3.1)$$

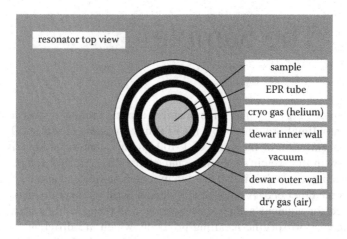

FIGURE 3.1 Maximal dimensions of a frozen aqueous sample. Top view cut through the middle of a rectangular X-band resonator with sample tube and cryogenic gas flow system in place, indicating maximum sample tube diameter.

The effective measuring area of the rectangular X-band TE_{102} cavity is in fact 15 mm in the vertical direction, and so the maximum sample volume is indeed 0.2 mL. We need some additional tolerance because we cannot see the sample when placing it in the resonator and, of course, we want to make full use of the measuring area not only for maximal signal intensity, but also for reproducibility. Increasing the sample size to 250 μL gives $h = 1.88$ cm, so almost 2 mm tolerance on top and bottom to "blindly" center the sample in the right position. In point of fact, since water expands by circa 10% in volume upon freezing, and the only direction in which it can actually expand in a hard tube is vertically, the actual tolerance will be closer to 3 mm on both ends. So 250 μL per sample is the ballpark number to remember when you get ready for the cold-room protein purification facility or, alternatively, for negotiations with your friendly colleague willing to act as supplier of interesting preparations. Top this request off with the adage to go for the "highest possible" sample concentration compliant with the solubility of the particular biomolecule and we will be in an optimal spectroscopic starting position. For example, most "water soluble" proteins are readily dissolved at least up to 50 mg/mL (in some cases up to 200 mg/mL), and for an averaged-size protein of 50 kDa this would give a concentration of 1 mM, which is generally high enough for the ready detection of even the broadest type of transition-ion spectrum. For many types of spectra concentrations of at least an order of magnitude less will actually be sufficient, which is fortunate because a total amount of 12.5 mg protein (i.e., 50 mg/mL in 250 μL) may not always be obtainable due to limiting amounts of available biomass.

What material is the sample tube made of? Just like a cuvette in a UV-visible spectrometer has to be optically transparent, the EPR sample tube must be transparent for the magnetic component of microwaves. High-quality quartz is the preferred construction material; low-quality quartz and especially any type of glass will not do, because these materials contain paramagnetic metal contaminants, in particular iron centers, with EPR spectra similar to those of iron proteins. The "paramagnetic" cleanness of quartz can not be ascertained by visual inspection; the only proper test is to measure its EPR spectrum.

What is the total length of a quartz EPR tube? The frozen 250 µL sample in a tube of 4 mm inner diameter has a height of 21 mm so the tube length should at least be greater than this number. The samples should not be oversized (significantly more than 2 cm) to minimize the risk of tube cracking during freezing or thawing (see below). The dimensions of the cooling system puts some constraints on the tube length: typically, a length of 20–25 cm is required to get the sample-containing end in the middle of the resonator with the other end sticking out of the dewar in the air at ambient temperature. However, a tube of 25 cm is relatively fragile and cumbersome to handle, for example, when connected to a vacuum/gas manifold, and filling it with sample requires needles of uncommon length. Furthermore, when stored in liquid-nitrogen containers or in −80°C freezers, long tubes take up a lot of space. A length of 13–14 cm is a good compromise. It is a practical size for handling in the wet lab. When the sample is in place in the spectrometer, the other end still sticks out of the X-band resonator. Many tubes can be cryo-stored in typical drawers and/or boxes in common use in molecular biology. For use in a cooling system the tube probably has to be fitted with an elongation rod, for example, made of hollow stainless steel or of solid plastic as illustrated in Figure 3.2. The steel rod may clamp around the quartz tube; the connection with the plastic rod can be made of soft rubber or, when tolerance is limiting (i.e., when the rubber could block the flow of cryogenic gasses) the brown tape in use by the postal services is a good alternative.

For proper bookkeeping and archiving each sample tube must be uniquely labeled. A label such as "brownish fraction number 73 from DEAE column, possibly containing a new iron–sulfur protein, concentrated over Amicon 30, reduced

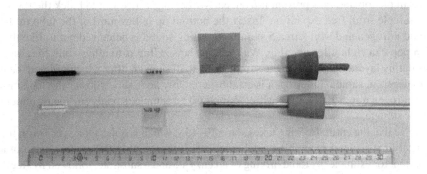

FIGURE 3.2 The sample tube and its peripherals. Two low-temperature X-band sample tubes of circa 13 cm length and 0.5 cm diameter and their elongation rods are shown. The upper tube has a wrapped label (W99) and holds a colored sample (to be frozen) of 3 cm height; it is ready to be connected to its elongation tube by means of a 3 × 3 cm piece of tape. The lower tube holds a colorless sample of 2 cm height; its 1 × 1 cm label (W98) is still unwrapped. It will be connected to a clamping metal elongation rod. The corks are to fit the helium-flow dewar in Figure 2.10.

with 2 mM dithionite, in 50 mM Tris-HCl, pH 8.0" may be informative and unique, but unfortunately it doesn't really fit on a 13 cm tube in a readable font. We use a short alphanumerical alias and take care of proper documentation in a lab journal, for example, WH374 for operator WH working on one of his samples in the range 001–999. The label is written with a black ballpoint pen on a square cm of yellow tape (used for autoclaving; also used by housepainters for window lining). This tape remains attached to the tube after repeated immersion in liquid nitrogen, which is not true for regular white paper labels, or for cello tape. Writing from blue ballpoint pen, crayon, or fine liner may rapidly fade in liquid nitrogen. The square cm size of the tape is to get one single overlapping wrap around the quartz tube; more tape increases the tube diameter and may block the flow of cryogenic gas. The label is attached 1 cm below the top of the tube to leave space for attachment of the elonga-tion tube (Figure 3.2). Also, after a sample has been permanently eliminated from the storage, the yellow tape is relatively easily removed, and the tube can be washed and reused. Alternatively, tubes may be permanently labeled by engraving or by in-burning a special label.

3.2   FREEZING AND THAWING

Some care is required when freezing or thawing aqueous EPR samples in liquid nitro-gen. Water has the exceptional property to expand upon freezing and the released force may easily crack EPR tubes. Chances for this to happen are roughly one out of two when a tube with a sample is dumped into liquid nitrogen. Such a disastrous failure rate is easily circumvented by *slowly* freezing the sample from the bottom to the top so that the expansion is always in the upwards vertical direction towards the tube opening. Here is the fault-proof recipe: take a (small) open dewar and fill it to the rim with liquid nitrogen (Figure 3.3). Grab a sample tube between thumb and index finger close to the top of the sample; this will provide some body heat to the sample top and thus prevent it from early freezing, which would block the rest of the sample from free expansion. Insert the bottom tip (a few mm) of the tube in the liquid nitrogen and *wait* (circa 5 s) until a sizzling sound is perceived not unlike that of a pop can right after opening. Waiting for the sizzling is to allow time for a thin insulating layer of gas around the sample tube to disappear, after which efficient heat transfer from sample to nitrogen is established. Only now, slowly lower the tube into the nitrogen (circa 10 s for the whole sample). The sample is ready for measurement or for storage.

Note that the entire freezing takes some 10–15 s, which is quite a long period on the typical time scale of many biological reactions. If you want to follow the course of a reaction, for example, by rapid mixing of an enzyme and a substrate, followed by freez-ing for EPR spectroscopy, the "dead time" of the overall procedure is at least 10–15 s. This long freezing time can also cause differential freezing rates over the sample and therefore sample inhomogeneity, i.e., concentration gradients and local clustering of molecules, which is undesirable for one because the spectra will be interpreted under the assumption of homogeneity and random orientation. For inhomogeneity to occur is relatively rare with proteins, but not so rare with small coordination compounds. A notorious example is the $Cu^{II}(H_2O)_6$ complex that we use as a convenient external

FIGURE 3.3 Fail-safe recipe for freezing EPR samples. The tube is held just above the sample, which is allowed to freeze and expand from the bottom to the top.

$S = 1/2$ EPR concentration standard. The complete recipe for the standard is 1 mM $CuSO_4$ + 10 mM HCl + 2 M $NaClO_4$. The hydrated cupric ions are avid to di- and polymerize, which is strongly suppressed by the addition of a large amount of the "non-ligand" perchlorate. Even with this precaution some inhomogeneity may still occur, but now it is easily cured as follows: Directly after the sample has been frozen in liquid nitrogen, take it out again and hold it in air for half a minute, and then dump it back in the nitrogen. During the half minute the sample warms up from 77 K to above 175 K, which is enough for a second-order phase transition visible as the disappearance of ice cracks and the evolution of a homogeneous (milky) sample appearance. Now the sample *is* homogeneous as can be seen by comparing the spectra of $Cu^{II}(H_2O)_6$ before and after "warming" in Figure 3.4. If a freezing time of 10–15 s is unacceptable, either because of inhomogeneity problems or because one is interested in pre-steady-state reaction rates on a faster time scale, the liquid nitrogen can be replaced with a cryogenic of higher heat capacity and more efficient heat transfer. The most widely used coolant is the organic solvent isopentane with a freezing point of −160°C (= 113 K).

At some point in time we may want to thaw the sample, for example, because we want to discard it, or because we want to use it for other experiments, or because we want to add a reactant (e.g., substrate or reductant) before freezing it again for

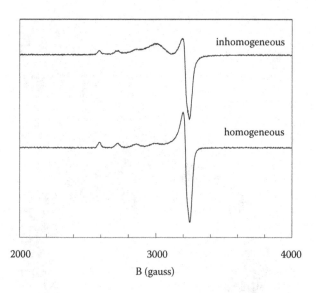

2000   3000   4000

B (gauss)

FIGURE 3.4 Spectral evidence for freezing-induced sample inhomogeneity. A solution containing 10 mM $CuSO_4$ + 10 mM HCl + 2 M $NaClO_4$ is frequently used as external $S = 1/2$ concentration standard. The paramagnet is the coordination complex $Cu^{II}(H_2O)_6$ which has the tendency to polymerize during freezing (spectrum A); warming up to circa 200 K leads to homogenization (spectrum B).

further EPR studies. Once more, we must have a procedure to prevent the tube from cracking during the handling. Thawing is not only the physical opposite of freezing; also, our thawing procedure is, to some extent, the opposite of our freezing procedure: freezing should be slow, thawing should be fast, but also here the details are important. When a frozen sample is taken out of the liquid nitrogen and left on the bench to slowly thaw, the risk of tube cracking is roughly one out of ten. A fail-safe recipe is as follows: Fill a beaker of circa 100 mL volume two thirds with lukewarm tap water. Take the sample out of the liquid nitrogen and hold it in air in a vertical position at some distance from your eyes, and wait (circa 15 s) for it to "erupt," that is, for any air to blow off that has previously condensed to liquid in the tube on top of the sample. Then, put the sample tube in the lukewarm water so that the sample itself is completely immersed below water level, and immediately start stirring with the tube until the sample has thawed. Take out the tube and wipe off the exterior. The sample is now ready for further processing.

Nonbiological, synthetic model compounds may not be soluble in aqueous solution. In contrast to water, other solvents do not expand upon freezing but rather slightly shrink. Consequently, the freeze-thaw procedures for aqueous samples do not apply here. Tubes with samples in nonaqueous solution may be rapidly frozen by direct immersion in liquid nitrogen without further precautions. Contrarily, thawing has to proceed slowly. Take a small, open dewar containing a small amount of liquid nitrogen. Throw out the nitrogen, place the tube with the frozen nonaqueous solution into the empty cold dewar, and wait until the sample has thawed completely. The sample is now ready for further handling.

3.3 SOLID AIR PROBLEM

A key characteristic of Nessie, the monster of Loch Ness, is that although one may be convinced to have accidentally spotted it in the past, whenever one tries to actually characterize the beast, it will categorically refuse to give any sign of existence. EPR spectroscopy has its own equivalent of Nessie, and its name is solid oxygen. Being uninformed of its apparently erratic behavior can easily cost you several months of your productive scientific life, so read on to be prepared for the inevitable encounter.

The ground state of O_2, molecular oxygen, is a triplet, which means that it is a paramagnetic compound with two unpaired electrons and $S = 1$. The solid state of this molecule is the monster. The X-band spectrum of gaseous and liquid oxygen consists of many lines at fields above circa 4000 gauss (Beringer and Castle 1951), which usually do not interfere with our spectroscopy because they are of low intensity, for example, due to pressure broadening in atmospheric air. Pure oxygen becomes solid at 54.8 K, but it does not have an EPR spectrum due to the formation of an "antiferromagnetic" lattice, that is, the individual O_2 magnets are oriented in such a way as to completely cancel each other's magnetism. Not so for oxygen in air (21%). Instead of the 14 cm standard sample tube, let us take a longer (say, 25 cm) tube and place it empty in the spectrometer with its opening in the air. When liquid-helium cooling is applied, the air in the open tube will condense at circa 79 K and then solidify at circa 60 K. Figure 3.5 is the X-band spectrum of such a sample taken at 4.2 K: the whole magnetic field range is covered with lines. The figure also gives the spectrum taken in parallel-mode configuration; this technique will be explained in the next and later chapters. When the temperature is raised above 4.2 K, the positions and shapes of these lines change in a complex manner. Above circa 25 K the spectrum disappears altogether.

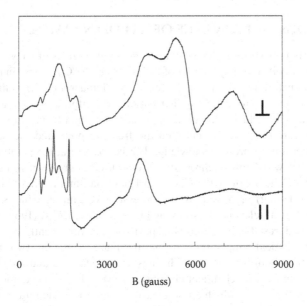

FIGURE 3.5 Solid-air EPR. The regular (9.65 GHz) and the parallel-mode (9.36 GHz) X-band spectrum of solidified air at 4.2 K.

When a regular EPR sample tube is taken out of a storage dewar to be put into the spectrometer, on its way it may condense air on its outside (and inside on top of the sample), and this air will become solid when the sample is cooled in the spectrometer to temperatures below 25 K. Many metalloproteins have their optimal EPR measuring temperature at <25 K, and their spectra will contain the solid-oxygen lines. Because of the sensitivity of the shape of the oxygen spectrum to the temperature, it is not always easy to recognize it. I have personally witnessed more than half a dozen instances in which experimentalists (biochemists and physicists alike) were convinced to have discovered an excitingly novel and complex EPR signature from their samples, which unfortunately would come and go, as the Loch Ness monster, and which eventually (in several cases after many months) would disappear completely once the experimentalist would systematically take the precaution to keep the sample tube in air for a few seconds, then to wipe it off with a cotton cloth, and only then to place it in the spectrometer.

In recent years O_2-Nessie has also appeared in high-frequency (≥ 95 GHz) EPR and its spectroscopists have given the encounter a positive twist by adopting the monster as a spectroscopic model system: its analysis at high frequency turns out to be straightforward and an illustrative application of high-frequency EPR to concentrated high-spin systems (Pardi et al. 2000). Interestingly, the high-frequency analysis does not in any way predict the X-band spectrum of Figure 3.5 at all. Moreover, 65 years of EPR literature has not even afforded a single allusion, be it experimental or theoretical, to this elusive spectrum.

The bottom line: whenever you find yourself in the state of exaltation that comes with the putative discovery of a novel and unusual signal, count to ten, wipe the frozen air off the tube, remeasure it, and be prepared to find that your discovery has evaporated.

3.4 BIOLOGICAL RELEVANCE OF A FROZEN SAMPLE

Biological activity is strongly dependent on temperature; rates of enzyme-catalyzed reactions typically increase by a factor of 2–3 for every 10°C increase in temperature until protein denaturation sets in. More generally, biomacromolecules may occur in different conformations (or different distributions of conformations) as a function of temperature. By the exclusive choice of water as its solvent, life has limited its operational temperature range to fall between the freezing point and the boiling point of water. According to present knowledge life is possible at temperatures not less than slightly below 0°C of psychrophilic (= cold loving) organisms in supercooled Antarctic waters, and not more than the circa 121°C maximum of hyperthermophilic (extremely hot loving) organisms in high-pressure volcanic marine niches. What then is the biological relevance of studying biomolecules at 77 K (196°C) or below?

Let us first address the question why spectroscopists frequently have to employ these nonphysiological temperatures in the first place. There are two fundamental reasons, and they both have to do with rates: a physical rate and a chemical rate. The physical rate (i.e., no chemical changes involved) is the relaxation rate of the paramagnet, or the rate at which a microwave-excited molecule returns (falls back) to its ground state. This rate increases with increasing temperature. High paramagnetic relaxation rates cause broad lines; very high relaxation rates cause lines to be broadened beyond detectability. Many paramagnetic transition ions in proteins have such a high relaxation rate that their EPR spectra are too broad to be measured at ambient temperatures. We must cool the samples often to temperatures close to the boiling point of helium to get a signal at all. We will return to this theme in Chapter 4. The chemical rate is, of course, the rate at which a sample proceeds to react following its preparation in the EPR tube. If we deliberately want to avoid this reactivity as, for example, in the case of an unstable, rapidly decaying radical, then we must freeze the reaction mixture to stop (or at least very drastically slow down) the reaction rate, so that we can leisurely take an EPR snapshot of the radical.

The next question to address is to what extent does the study of a (deeply) frozen biological sample provide information that is relevant for an understanding of its functioning in a living cell at whatever the "ambient" temperature of this cell happens to be? First and foremost, let us state the fact of experience that solutions of biomacromolecules such as metalloproteins can be frozen and thawed many times without any detectable deterioration of their biological activity. Combined with the rather low intensity ($\leq$0.2 W) of the microwave source of an EPR spectrometer, this leads to the proposition that EPR spectroscopy is a nondestructive technique.

The reason why I call this a "proposition" is that it is not a law of nature carved in stone. Occasionally, one may come across proteins that are cold labile, especially complex ones associated with biological membranes such as the ATP-synthase complex. More generally, freezing/thawing of any biological preparation, whose functioning depends on the intactness of an ion-impermeable membrane, will lead to rupture of the membrane and loss of function unless specific precautions are taken such as the addition of "antifreeze." For example, a single freeze/thaw cycle of so-called "coupled" mitochondria (i.e., in which oxidation of NADH is coupled to the phosphorylation of ADP to ATP via the generation of a proton motive force) results in complete uncoupling (i.e., after thawing NADH is oxidized without ATP production). However, if the mitochondria are incubated in 10% glycerol prior to freezing, then circa 90% of oxidative phosphorylation activity is recovered after a single freeze/thaw cycle.

Anyway, purified, soluble proteins are presumably stable towards repeated freezing and thawing cycles. Definite data on the subject are hard to find in the literature, which is perhaps understandable: An unnamed Ph.D. student once spent three boring days of his existence by freezing and thawing a solution of horse heart cytochrome c for 128 subsequent times while regularly checking the fine details of the low-temperature EPR spectrum for any changes however small. The outcome of this experiment was no detectable spectral changes whatsoever, which is gratifying for those who encourage the promotion of bioEPR spectroscopy, but what editor of a respectable journal would be willing to publish such a trivial nonresult?

Having ascertained that low-temperature EPR is nondestructive, we can now address a fundamental follow-up question: To what extent does the spectroscopy of a biomolecule at a temperature of, say, 10 K bear relevance to that same molecule's cellular functioning at its ambient temperature of, say, 310 K (=37°C). More specifically, to what aqueous solution conformation temperature, if any, does a frozen solution protein conformation correspond? This now is a really hard question to answer, simply because it is difficult to approach experimentally, and consequently, there is little relevant data in the literature. We must identify a monitoring parameter that can be probed both in the liquid and in the solid state. The EPR spectrum itself will not do because at room temperature it is undetectable or severely broadened. The reduction potential, E^0, of redox proteins could be a relevant parameter. Reduction-potential values are a reflection of the relative stabilities of the oxidized and the reduced form of a redox couple, and these stabilities are a function of the temperature. Using EPR spectroscopy as the monitor for the concentration of one of the two forms of the redox couple, the reduction potential can be determined using sample points drawn from a bulk solution titrated with a chemical reductant or oxidant with constant measurement of the solution potential (experimental details of this procedure will be discussed in Chapter 13). For some redox proteins—in particular, for small electron-transfer proteins—the E^0 can also be determined by direct voltammetry on a carbon electrode using a potentiostat, and this determination can be done as a function of solution temperature. Figure 3.6 gives the result of such a combined experiment for the small ET (electron transfer) protein rubredoxin purified from a hyperthermophilic organism (Hagedoorn et al. 1998). Rubredoxin contains a single Fe(III/II) ion tetrahedrally coordinated by four cysteinato ligands. The protein has prolonged thermal stability up to the boiling point of water, and so the experiment can be carried out over a relatively wide range of temperatures. Using voltammetry a linear dependence of E^0 on T is found over a wide range of 25–90°C with a slope of circa −1 mV per degree which is a common value for ET proteins. Contrarily, using EPR monitored bulk titration, the original temperature of the solution is seen to be irrelevant: for all temperatures the same E^0 is found and this constant value corresponds to an extrapolated E^0 from the voltammetric experiment at a temperature of

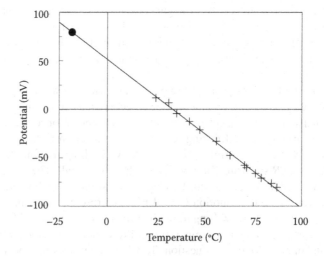

FIGURE 3.6 Determination of the reduction potential of rubredoxin by electrochemistry and by EPR monitored bulk titration. The (+) data are from cyclic voltammograms taken at different temperatures; the (•) point is from low temperature EPR monitored titrations at ambient temperature or at 80°C. (Data from Hagedoorn et al. 1998.)

circa −18°C. In other words, since for the EPR measurement the sample has to be frozen in liquid nitrogen, we always obtain the E^0 that corresponds to the approximate freezing temperature of the protein solution independent of the original temperature of the titration experiment. This interpretation implies that the rate of conformational adjustment of the molecular structures of oxidized and reduced rubredoxin with changing temperature is faster than the dead time of the freezing experiment even if we freeze 80°C samples rapidly in isopentane of −160°C.

Are biomacromolecules of greater mass than that of rubredoxin (6 kDa), in particular enzymes (typically, ≥50 kDa), capable of such rapid conformational adjustment with decreasing temperature? At present the answer appears to be: We do not know. Unfortunately, the reduction potential(s) of enzymes in solution is not usually determinable with direct electrochemistry, so you are invited to find and explore other molecular properties to probe as a function of temperature, for example, (de) protonations near paramagnetic sites that can be followed both by optical and by EPR spectroscopy.

3.5   SAMPLE PREPARATION ON THE VACUUM/GAS MANIFOLD

Redox-chemical considerations are innate to the mental framework of the bioEPR spectroscopist. Transition metal ions by their very nature can occur in two or more oxidation states, and their ground state can either be diamagnetic (i.e., no EPR), or paramagnetic with an even number of electrons (i.e., with difficult EPR), or paramagnetic with an odd number of electrons (i.e., with easy EPR). We will discuss the relation between spin state and "ease of EPR" later in detail; for now we state that it is very useful to explore redox transitions, for example, in order to be able to change from an oxidation state without an EPR spectrum to one with a spectrum. To this goal the first condition to be fulfilled is the efficient exclusion of air because it consists for circa 21% of the relatively strong and reactive oxidant O_2. Secondly, our bioEPR survival kit should contain a number of redox chemicals that are of general applicability in biochemistry, that is, that are sufficiently strong reductant or oxidant to be capable of reducing or oxidizing a large range of biomolecules without leading to any deterioration, denaturation, or destruction of those biomolecules.

Once more we recall that water is the solvent of life, and we furthermore appreciate the fact that life usually tries to avoid extremes in acidity or basicity and, therefore, that physiological activity usually thrives at pH values in the neighborhood of that of neutral water (even extremophilic acidophiles of alkaliphiles have an internal cellular pH near neutrality). This implies that the extremes in biological redox chemistry are expected to be found at potentials corresponding to the reductive decomposition of water (hydrogen evolution) and the oxidative decomposition of water (oxygen evolution). By definition all electrochemical potentials are relative to that of a model couple, namely, the $H_2/2H^+$ couple of the normal hydrogen electrode (NHE) at pH = 0 (1 M H^+) and partial pressure p_{H2} = 1 atm with a standard reduction potential $E^0 \equiv 0$ volt. From the Nernst equation for a compound X:

$$E = E^0 + (RT / nF)\ln([X_{ox}] / [X_{red}]) \qquad (3.2)$$

which at ambient temperature (25°C) can be written as

$$E \approx E^0 + (0.0591 / n) \log([X_{ox}] / [X_{red}])$$ (3.3)

in which the universal gas constant R = 8.3145 J K^{-1} mol^{-1}, T is the temperature in kelvins, n is the number of electrons transferred, and the Faraday constant F = 96,485 J V^{-1} mol^{-1}, it follows that the $E_7^0(H_2/2H^+)$ is seven times 59.1 millivolts more negative than the NHE value, and so this potential of −414 mV is the reference number for biological redox chemistry. This defines an approximate working window of "allowed" (i.e., nondestructive) redox potentials as illustrated in Figure 3.7. The oxidative limit of the $2H_2O/O_2$ couple is also indicated, and so are the reduction potentials under biochemical standard conditions, E_7^0 (or E'^0; or, frequently, in biochemical literature $E_{m,7}$ or simply E_m with subscript m for midpoint) of a number of general reductants and oxidants commonly applied in biochemistry. The gasses H_2 and O_2 themselves could have been ideal general reductant and oxidant were it not for the fact that actually very few biomolecules react with hydrogen and that, although there is a significantly higher number of biomolecules that do react with oxygen, unfortunately many are, in fact, destroyed in the action. In short, we have to employ other chemicals and, furthermore, since oxygen frequently reacts rather vigorously with the chemical reductants, we have to be able to rigorously exclude O_2 from our EPR samples. To this goal the vacuum/gas manifold or Schlenk line is an absolutely essential labware tool. Its outline is drawn in Figure 3.8. A number of three-way stopcock glass valves (or cross valves) are arranged in such a way that one arm of each valve is outlet to a vacuum pump, a second arm is inlet for a pressurized

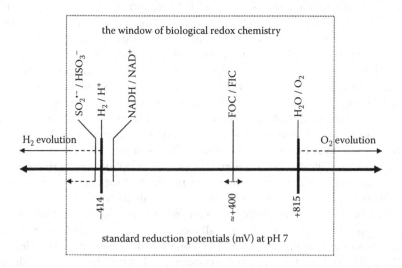

FIGURE 3.7 The potential window for the redox chemistry of life. Redox chemistry in living cells is approximately limited by the standard potentials for reduction and oxidation of the solvent water at neutral pH. Approximate standard reduction potentials are also indicated for the commonly used oxidant ferricyanide and reductants NADH and dithionite.

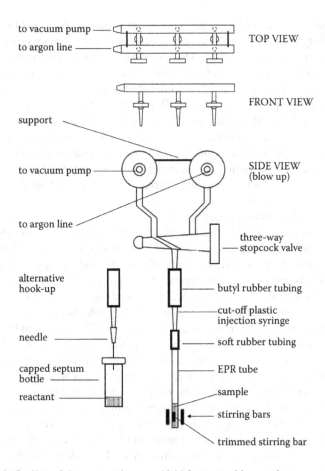

to vacuum pump
to argon line
TOP VIEW

FRONT VIEW

support

to vacuum pump
to argon line

SIDE VIEW
(blow up)

three-way
stopcock valve

alternative
hook-up

butyl rubber tubing

cut-off plastic
injection syringe

needle

soft rubber tubing

capped septum
bottle

EPR tube

reactant

sample

stirring bars

trimmed stirring bar

FIGURE 3.8 Outline of the vacuum/gas manifold for anaerobic sample preparation.

gas cylinder (with the pressure reduction valve opened to *not more than* 0.1–0.2 atm overpressure), and the third arm is the in/out connection to EPR tubes.

An EPR sample is made anaerobic by pumping away the air and replacing it with an inert gas, for example, nitrogen or, better still, argon. However, this is not at all a trivial act; directly pumping on a sample in an EPR tube will cause this sample to be sucked out of its tube and to end up in the waste of the vacuum pump. To avoid losing our precious samples this way, we first freeze them in liquid nitrogen, then we take them out of the nitrogen and wait a few seconds for any condensed air to evaporate, and then we hook up the tube, with the sample still frozen, to a valve of the manifold as illustrated in Figure 3.8. Immediately after connecting the tube the valve is opened to the vacuum line so that the pump will be working on the frozen sample. After 10–20 seconds, the cross valve is switched to argon (or any other inert gas available) for a few seconds. The vacuum/argon cycle (15/5 s) is repeated until the very top of the sample begins to thaw as evidenced by the onset of foam formation.

At this point in time the line is permanently switched to argon, and the sample is allowed to fully thaw. The sample is now anaerobic and is ready for incubation with an anaerobic reductant or oxidant. The anaerobic reactant solution is prepared as follows. A septum-sealed bottle of appropriate size (e.g., 2–5 mL), and with a small magnetic stirring bar inside, is filled with buffer solution, closed, and a gas-tight injection needle of appropriate size (e.g., Hamilton 25 µL), fitted with an elongation needle, is pierced through the septum. The bottle is now hooked up to the manifold (no freezing) as illustrated in Figure 3.8, and subjected to a few vacuum/argon cycles. Then, the lid is opened under argon, the solid redox reactant is added to the liquid, the lid is closed, and the vacuum/argon cycling is continued (5–10 cycles), while the stirring bar is operated with a magnetic stirrer to assist dissolution of reactant and evaporation of dissolved gas.

Finally, with the solution under argon the valve is closed, and the bottle is disconnected from the manifold. The injection needle is flushed with solution, filled, taken out, quickly transferred to the EPR tube on the argon line, pierced through the soft rubber connection, guided into the EPR sample, and the appropriate amount of reactant is discharged into the sample. The injection needle is now drawn out of the tube and the rubber connector, leaving a small hole in the rubber. The argon overpressure prevents air from entering through this hole, but note that application of vacuum has now become impossible. Mixing of the EPR sample and the reactant may be accomplished by means of a small magnetic stirring bar in the EPR tube (it may have been necessary to trim some of the Teflon coating of the bar with a nail file for it to fit in the EPR tube). Alternatively, mixing can be achieved by continuously finger-ticking the bottom tip of the tube with some force. Following an appropriate incubation time (e.g., 5–10 minutes for reaction with the sluggish reductant sodium dithionite), the internal stirring bar is lifted with a ring of external bars to the top of the EPR tube, and the sample is frozen in liquid nitrogen. Directly after freezing the argon line is closed, the EPR tube is disconnected, the pierced rubber connector is taken off and discarded, and a stirring bar is taken out of the tube. The sample is now ready for spectroscopy.

Oxygen-sensitive samples such as enzymes from strictly anaerobic microorganisms which have been previously prepared anaerobically (e.g., in a glove box) can be transferred from their container (e.g., a capped septum bottle) to an empty EPR tube that has been made anaerobic on the manifold. The transfer can be done with an injection needle that has been made anaerobic in an empty septum bottle on the manifold. Note that after the transfer the connecting rubber has a hole, so any addition to this sample can only be made after it has been frozen and subsequently made anaerobic with a new, intact connecting rubber on the manifold.

It is practical to place a "washing bottle" or "scrubber" in the gas line just before the manifold. The aqueous solution in this bottle contains a reductant for traces of molecular oxygen and at the same time "wets" the gas which will minimize a concentrating effect on the sample by drying. A practical solution is 1 mM zinc acetate, 1 µM TMP (meso-tetra(N-methyl-4-pyridyl)porphine-tetra-tosylate), 100 mM Na$_2$EDTA, 100 mM Tris-HCl buffer at pH 10. The "porphyrin" complexates the Zn^{2+} and forms a light-sensitive compound that can be excited by near UV light from an 18 watt TL-tube.

3.6 CHOICE OF REACTANT

Perhaps the most appropriate reductant of a metalloprotein would be its natural substrate, however, there may be several reasons why this is not always a practical option: the substrate may not be known; the substrate may be difficult to handle in aqueous solution; its reduction potential may not be low enough to ensure quantitative reduction of prosthetic groups; or the metalloprotein may not be a redox enzyme altogether. For all these common situations we have a small number of general reductants and oxidants available so that a change of metal oxidation state is generally possible, and so application of EPR spectroscopy to biological transition ion complexes is generally practical.

Since the lower limit of biological redox chemistry is approximately at the potential of the hydrogen electrode, a general reductant should have an $E'^0 < -414$ mV. By far the most common nonphysiological reductant for biomolecules is sodium dithionite, $Na_2S_2O_4$, which has been found to be very widely applicable, and this has unfortunately also led to frequent misuse in the form of adding "a few grains" of solid dithionite to aerobic solutions. A few grains easily add up to molar concentrations and the reaction with oxygen leads to acidification, which may destroy the metalloprotein under study (e.g., many iron–sulfur proteins are unstable at pH ≤ 6 because they contain "acid-labile" sulfur). The actual reductant is not really dithionite: the anion splits (to a very small extent) into two $SO_2^{\cdot}$ radicals, and these are the actual reducing agent. It is important to realize that the apparent reduction potential of dithionite is strongly dependent both on the pH and on the dithionite concentration. The E^0 becomes less negative with decreasing pH and with increasing dithionite concentration. A colorless general-purpose reducing stock solution is circa 100 mM $Na_2S_2O_4$ in 1 M Tris SO_4, pH 8–9, which is anaerobically prepared on the manifold as described above, and which is anaerobically added to EPR samples of 200 μL volume to the amount of 10–20 μL affording a final concentration of 5–10 mM. This creates an effective reduction potential of circa −460 mV (Mayhew 1978). Using higher concentrations does not lead to lower final solution potentials.

The $E'^0 = -414$ mV value is not a hard lower limit for biological redox chemistry: several complex redox enzymes can catalyze reactions at even lower potential in their interior, and so they can have redox active metal prosthetic groups with $E'^0 \approx -0.5$ V or lower. If dithionite turns out to be just not good enough, one can try the alternative of Ti(III) citrate with $E'^0 \approx -500$ mV. The solution is typically prepared as a black 100 mM stock in a glove box from an ampoule of Ti(III)Cl$_3$ and sodium citrate titrated with Na_2CO_3 to pH 8–9 (Holliger et al. 1992). A distinct disadvantage is the fact that Ti(III) is a d^1 ion and therefore an $S = 1/2$ paramagnet itself, and the citrate complex exhibits an EPR spectrum that strongly overlaps with, for example, those of iron-sulfur clusters (see Figure 3.9). At lower pH values the compound dimerizes with an $S = 1$ triplet ground state, which has a "half field" signal around $g \approx 4$.

If everything else fails, then the ultimate reductant is the flavin/light system (Mayhew 1978). Deazaflavin is a generic name for derivatives of the isoalloxazine ring system lacking the 5N, for example, 3,10-dimethyl-5-deaza-isoalloxazine. These compounds can be excited by white light, for example, from a tungsten bulb using a light pipe. In the light-excited state they are strong oxidants capable of destructively

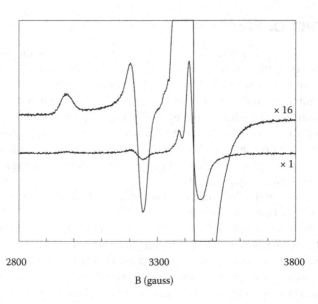

2800   3300   3800

B (gauss)

FIGURE 3.9 EPR of titanium citrate. The citrate complex of the Ti(III) ion at pH 9 is a general-purpose strong reductant of metalloproteins. This $3d^1$ system gives an $S = 1/2$ EPR spectrum with g-values just below g_e. The axial signal below 3300 gauss is from Ni(I) in factor F-430.

oxidizing EDTA. The reduced deazaflavin when fallen back to the ground state is a strong reductant ($E'^0 \approx -1$ V), and in the presence of redox mediator methyl viologen (E'^0's of -450 mV and -650 mV) not many metalloproteins can escape reduction. A practical drawback of the system is that deazaflavin compounds have to be synthesized as they are not broadly commercially available. An alternative is the cheap compound proflavin, which, however, reacts more sluggishly in the light/EDTA reaction. Degree of reduction of metalloproteins can be followed by EPR, and after each measurement the sample can be thawed (anaerobically!) and further illuminated.

Quantitative oxidation of biological metal centers is usually less of an issue, where many metalloproteins are found to be fully oxidized upon isolation and purification. This does, however, not generally hold for anaerobically purified proteins (which may become damaged upon exposure to oxygen), and it is also not always true for centers with multiple oxidation states (Co, Mo, W, metal clusters, flavin). For these systems a practical general purpose oxidant is the ferricyanide (oxidized hexacyanoferrate) anion from $K_3Fe(CN)_6$. Reduction by outer sphere electron transfer to the ferrocyanide anion $[Fe(CN)_6]^{2+}$ has a standard reduction potential at pH 7 with a significant ionic-strength dependence (Hanania et al. 1967) which is $E'^0 \approx +400$ mV at the typical ionic strength of circa 0.1 M of a buffered protein EPR sample. The $3d^5$, low-spin Fe(III) complex is an $S = 1/2$ system with an EPR spectrum (cf. Figure 3.10) similar to that of low-spin hemoproteins, but with such a large linewidth that up to millimolar concentration of ferricyanide spectral interference can be dealt with by construction of difference spectra.

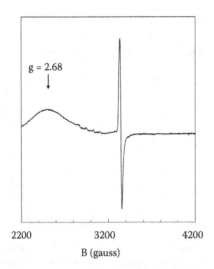

FIGURE 3.10 EPR of ferricyanide. Potassium ferricyanide is a general-purpose oxidant of metalloproteins. The low-spin $3d^5$ Fe(III) of frozen dissolved $K_3Fe(CN)_6$ gives a broad $S = 1/2$ spectrum with main peak at $g \approx 2.68$ ($\nu = 9407$ MHz; $T = 15.5$ K). The sharper feature around $g \approx 2$ is from an iron–sulfur cluster anaerobically oxidized in this experiment by ferricyanide to the $[3Fe-4S]^{1+}$ form with $S = 1/2$.

Some proteins may have redox centers with E'^0's significantly higher than +0.4 V (e.g., the multicopper protein laccase). No general approach is presently available to deal with the high end of the biological redox window, and, in fact, the reduction potentials of these systems (again, e.g., laccases) are frequently poorly defined.

3.7 GASEOUS SUBSTRATES

For oxidation by atmospheric oxygen a sample in an EPR tube may simply be opened to air and stirred. However, reaction with any other gas than air requires special handling of the sample on the manifold of a Schlenk line. Examples are oxidation by pure oxygen, reduction by hydrogen, and also the reaction by any gaseous substrate or inhibitor such as CO, CO_2, NO, N_2O, etc. Basically, there are two different experimental approaches: mixing with a solution in which the gas is dissolved or mixing with a pressurized atmosphere of the gas.

The first method is perhaps experimentally the easiest, but it is limited by the solubility of the gas. A septum-capped bottle with pure water, or with the buffer solution that is also used for the metalloprotein, is purged with the desired gas to saturation. Note that some gasses react with oxygen (e.g., NO) which means that the bottle first has to be made anaerobic by purging with an inert gas. An elongated gas-tight Hamilton injection needle is then used to anaerobically transfer an amount of the saturated-gas solution to the (anaerobic) EPR sample that is waiting on the manifold. Mixing of the protein sample and the dissolved-gas sample is effected by repeated refilling and discharging the Hamilton while disturbance of the solvent-gas meniscus is avoided

TABLE 3.1
Solubility (mM) of gasses in water at 0°C and 25°C

Oxygen	2.20	1.27
Oxygen in air (20.95%)	0.46	0.27
Hydrogen	0.98	0.79
Nitric oxide	3.32	1.94
Carbon monoxide	1.69	0.99
Nitrogen	1.07	0.65
Nitrous oxide	60.2	24.7
Carbon dioxide	80.0	34.1
Methane	2.59	1.41

to prevent exchange with the inert gas above the EPR sample, which would lead to dilution of the reactant gas. This approach affords semiquantitative addition of gas reactant, however, most gasses of biological interest have a solubility of circa 1 mM in water at ambient temperatures and so obtaining sufficiently high concentration of substrate may imply considerable dilution of the EPR sample, or may simply be impossible. A practical trick to increase dissolved gas concentration roughly by a factor of 0.7 is to keep the solvent (and also the EPR sample) at 0°C on an ice bath (see Table 3.1).

In case this concentration is still insufficient, the sample has to be reacted by direct application of the pressurized gas with vigorous stirring through the liquid–gas meniscus. To this goal a pressurized bottle of the desired gas has to be connected to the manifold, for example, as a replacement of the inert gas. Note also that some gasses require specific scrubbing, for example, NO is usually contaminated with NO_2, which has to be eliminated by leading the gas through a washing bottle with 2 M KOH.

3.8 LIQUID SAMPLES

From the point of view of biological relevance ideally EPR spectra should be taken from aqueous solution samples at physiological temperatures. Not-so-ideal reality brings along two major practical problems: paramagnetic relaxation and dielectric absorption.

We have already seen that bioEPR is frequently done on frozen samples at cryogenic temperatures because paramagnetic relaxation (in particular spin-lattice relaxation; see Chapter 4) of transition ions is usually fast at ambient temperature leading to spectra that are severely broadened frequently beyond detection. Lowering of the temperature reduces the relaxation rate, and this leads to sharpening of the spectra. However, not all biologically relevant paramagnets are too rapidly relaxing for room-temperature EPR. For some $S = 1/2$ systems, notably Ni(I), Cu(II), Mo(V), and W(V), broadened spectra are observable in solution. Also, most members of the broad class of radicals with a single unpaired electron afford room-temperature spectra that are not broadened at all by spin-lattice relaxation. This class does not only encompass the natural radicals from, for example, flavins and quinones, but also the widely used synthetic spin labels (e.g., TEMPO) and spin traps (e.g., DMPO) to be discussed in Chapter 10.

Microwaves and water may form a hot cooking team in the microwave oven, but in spectroscopy we rather prefer our electromagnetic radiation not to completely disappear in our samples to emerge subsequently only as heat. Because of the high value of the dielectric constant of water at microwave frequencies particular care has to be taken to minimize sample occupancy in the resonator at positions of high electric-field density. There are two solutions to this problem. One is to reduce the diameter of the cylindrical sample, that is, to use a capillary sample tube. An extremely simple setup for routine measurements on large numbers of samples as, for example, amplitude measurements on a single type of spin-trap spectrum (as, for example, in light-induced radical studies on the shelf life of beers in translucent bottles) is as follows: a capillary is chosen such that its outer diameter fits into a tube normally used for frozen solutions. The capillary is fitted with a piece of rubber tubing ending in a stop cock. The tubing can be hand-pressed to suck up samples from a reaction mixture, and it also ensures that the capillary can be placed in the regular tube without touching its bottom (which would cause the sample to be drawn out of the capillary into the outer tube). After filling, the two concentric tubes are placed in the resonator, and the spectrometer is tuned. The capillary can now be removed, washed, refilled, and replaced in the regular tube without the need to retune the spectrometer, thus allowing a sample throughput rate that is only limited by the coordinative capacities of the operator.

An alternative solution to the dielectric water problem that applies in particular to standard TE_{102} rectangular resonators, is to use a "flat cell": a rectangular sample holder that is oversized in the vertical direction, and with one small cut-through dimension, say 1 mm, and the other dimension circa 10 mm, that is, approaching the bore of the cavity access holes (Figure 3.11). When in place in the resonator, this cell is rotated about its vertical axis to a position such that the 10 mm dimension points to a direction of minimal E-field.

3.9 NOTES ON SAFETY

Let us end this practical chapter on the preparation of bioEPR samples with a few important practical directives to keep you in good health:

- Always use safety goggles when handling liquid nitrogen in open dewars. Dewar vessels may break and then implode. Do *not* wear plastic gloves: they become brittle when in contact with liquid nitrogen, and the nitrogen may then leak through a crack, filling the glove and causing severe skin burns.
- Liquid cryogenic gasses like nitrogen or helium should always be handled in well ventilated places. When a liquefied gas container fails, for example, by a sudden leak due to mechanical damage, large amounts of gas may come free with the risk of asphyxiation and/or cold burns.
- Before thawing a frozen aqueous sample, allow a few seconds for condensed air to evaporate. Omitting the air eruption step and immediately putting the tube from the liquid nitrogen in the lukewarm water is very likely to lead to explosion of the tube due to rapid air expansion.

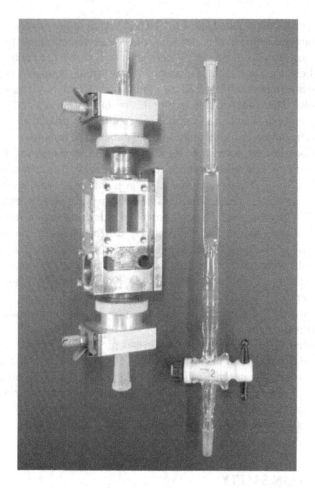

FIGURE 3.11 Aqueous-solution cells. Two different flat cells are shown; one is mounted with special clamps in a rectangular X-band resonator (the body has been opened for better view). Flat cells must be turned in the resonator to a position of minimal E_1-field.

- Removing a tube with a frozen *non*-aqueous solution from its liquid nitrogen storage, and then either holding it in air, or putting it on the bench, or dumping it in water, can lead to rapid thawing and therefore expansion of the sample resulting in explosion of the tube.
- Always use dedicated dewar vessels for handling liquid nitrogen on the bench. Do not ever use the food-grade thermos bottle in which you brought in your hot coffee. A thermos bottle can, in fact, hold liquid nitrogen, but its isolation is inferior to that of real dewars. This means that air is likely to condense to liquid in between the thermos bottle and its exterior plastic or metal cover. When you empty the thermos of nitrogen, the liquid air in the intermediate space will explode, and the exterior can be ripped open. Subsequently, the thermos itself may implode, and you will be showered with glass fragments.

4 Experimental Key Parameters

Low temperatures are required to slow down paramagnetic relaxation in order to get sharp EPR spectra. However, when a paramagnet can relax back to the ground state only slowly, then it is easy to saturate the system with microwaves, and this will lead to deformed spectra. In this chapter we consider the two key experimental parameters: power (intensity of the microwaves) and temperature (of the sample) in combination with the key system parameter: the spin. For a given system of spin S at a temperature T there is a single optimal value of P, which must be determined experimentally. The combined set of P, T, and S determines the complexity and the costs of EPR spectroscopy.

4.1 BOLTZMANN AND HEISENBERG DICTATE OPTIMAL (P,T) PAIRS

Consider a simple two energy level system, e.g., an $S = 1/2$ system. For a sample of realistic size at thermal equilibrium n_0 molecules are in the lowest energy state and n_1 in the highest state according to the Boltzmann distribution

$$n_1 = n_0 \exp(-\Delta E / kT) \qquad (4.1)$$

in which ΔE is the difference in energy between the two states (cf. Equation 2.3), $k = 0.695$ cm^{-1} K^{-1} is the Boltzmann constant, and T is the absolute temperature of the sample. The limiting values are $n_1 = 0$ (all molecules in the ground state) for $T = 0$ K, and $n_1 = n_0$ for infinitely high T. For X-band EPR $\Delta E \approx 0.3$ cm^{-1} and the population difference is rather small at laboratory temperatures as illustrated in Figure 4.1.

Detection of an EPR spectrum is only possible when microwave radiation can be absorbed by virtue of a finite difference in population of the energy levels involved ($n_1 < n_0$). Once the two populations have become equalized, the spin system is said to have attained an infinite spin temperature, and the spectrum should disappear. Because we know by experience that an EPR spectrum does not disappear upon repeated recording, there must be a way by which the spin system can get rid of the excess energy and can return to its original state of thermal equilibrium: spin-lattice relaxation. This term reflects the rooting of EPR in solid-state physics where the object of study is typically a crystal lattice containing a paramagnetic lattice defect. In a more general sense the lattice is anything that surrounds a paramagnet, for example, a protein surrounding a transition metal prosthetic group. In its turn the protein is coupled to a surrounding heat sink,

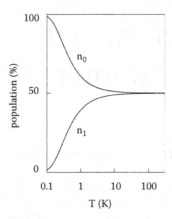

FIGURE 4.1 Boltzmann population of a doublet as a function of temperature. The lower and the higher level of the doublet have population n_0 and n_1, respectively, and they are separated by an energy difference of 0.3 cm^{-1}.

namely, a frozen aqueous solution in an EPR tube in, for example, a flow of cold helium gas. The energy transfer is by means of a coupling of spin transitions (relaxation to the ground state) with vibrations, or phonon transitions, in the surrounding "lattice." This can happen via three mechanisms that in the solid-state physics literature (e.g., Abragam and Bleaney 1970) are commonly referred to as (1) the direct mechanism (coupling of a spin transition to a single phonon of equal energy), (2) the Raman mechanism (coupling to a combination of any pair of phonons whose energy difference equals the energy of the spin transition), and (3) the Orbach mechanism (coupling with a specific pair of phonons via an excited spin level). These three mechanisms afford different dependencies on magnetic-field strength and on temperature, and they are furthermore functions of the spin S of the paramagnet and of the structure of the lattice. Quantitative data on relaxation in biological systems is very limited, but seem to suggest that the mentioned mechanisms are also operative in metalloproteins (Gayda et al. 1979). Obviously, a description of spin-lattice relaxation can become extremely involved, however, for almost all practical purposes of standard EPR on metalloproteins and models only two qualitative aspects are relevant: (1) the relaxation rate *de*creases with *de*creasing temperature and (2) the relaxation rate is anisotropic (i.e., is different for different parts of the spectrum).

The Heisenberg uncertainty principle dictates that energy and time associated with an atomic-scale system cannot be determined jointly within arbitrary precision:

$$\Delta E \Delta t > h / 2\pi \qquad (4.2)$$

For a two-level EPR system this reads as follows: when the life time of a molecule in the excited state is known accurately, then the energy of the excited state is uncertain. In other words, if spin-lattice relaxation from the excited state to the ground state would be infinitely fast, then the excited state life time would be exactly equal to zero seconds, and the uncertainty in the excited state energy would be maximal, which would lead to an EPR spectrum broadened beyond detection. Lowering the sample temperature decreases the relaxation rate, therefore increases the excited state lifetime, decreases the uncertainty in the excited state energy, decreases the EPR homogeneous linewidth, and increases the EPR amplitude.

This sharpening up of the spectrum by cooling the sample is, however, limited by a temperature-*in*dependent process: *in*homogeneous broadening. The protein or model molecules in dilute frozen solutions are subject to a statistical distribution in conformations. They each have slightly different 3-D structures and, therefore, slightly different *g* values, which manifest themselves as a constant broadening of the EPR line independent of the temperature as long as the medium (in casu ice) does not go through a phase transition. An $S = 1/2$ example (cytochrome *a*) is given in Figure 4.2.

Optimal (T,P) values for the recording of a metalloprotein EPR spectrum are determined by the interplay of all the above mentioned phenomena: the Boltzmann distribution, the Heisenberg uncertainty relation, the spin-lattice relaxation rate, and the conformational distribution of molecular structure. A practical optimization approach is to take data, using an arbitrary, intermediate value of P (e.g., circa 10 mW), at subsequently lower values of T until the spectrum does not sharpen up any further. Then, at this temperature, T_M (Figure 4.2), construct a normalized (with respect to electronic amplification: gain) power plot, as in Figure 4.3, to determine the maximum P value, P_M, of the linear range of the signal amplitude versus power, and the value $P_{0.5}$ required for half saturation. The subscript capital M can be read

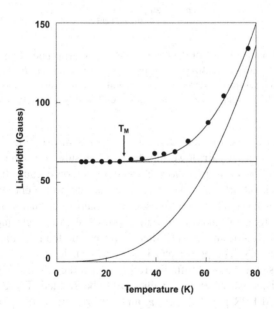

FIGURE 4.2 Linewidth increase with temperature for an $S = 1/2$ system. The linewidth of a feature in the low-spin heme spectrum from cytochrome *a* in bovine heart cytochrome oxidase has been fit as a convolution of a constant component from inhomogeneous broadening and a temperature-dependent component from homogeneous broadening (Hagen 2006). (Reproduced by permission of The Royal Society of Chemistry.)

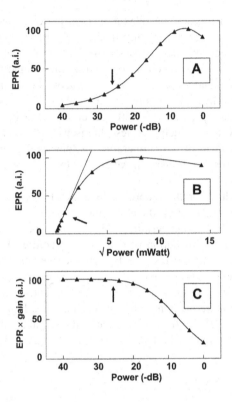

FIGURE 4.3 Alternative forms of the power plot. (A) experimental power plot; (B) linearized power plot; (C) gain-normalized power plot. The arrows indicate P_M: the limiting power value beyond which saturation sets in. Experimental data are from a [2Fe-2S]$^+$ Rieske-type protein (Hagen 2006). (Reproduced by permission of The Royal Society of Chemistry.)

as "measurement," which is short-hand for "optimal measurement condition." T_M and P_M form an optimal pair: lowering the power results in poorer signal-to-noise ratios; increasing the power leads to spectral deformation because the energy drain by spin-lattice relaxation cannot keep up with the microwave energy supply to the system. At power levels above P_M the system is said to be increasingly saturated. If P_M is greater than the maximum output power of the spectrometer (i.e., if *no* saturation is detected at any power), then lowering the temperature affords a real increase in sensitivity. The power plot is a characteristic of the molecular system, and it may be used as an—admittedly, rather unspecific—fingerprint. In quantitative EPR analysis, however, $P > P_M$ values are to be avoided. There is no theory to describe saturated EPR powder spectra, and determinations of spin concentration (see below) will be in error.

We have previously defined the relative dB scale in Equation 2.11. The power in EPR is expressed in decibels (dB) attenuation (or alternatively in −dB amplification) of a maximum value. X-band microwave sources (either klystrons or Gunn diodes) have a constant output that is usually leveled off at 200 mW. This value then corresponds to 0 dB attenuation. The conversion between absolute (watt) and relative (dB) power levels for a spectrometer with 0.2 W maximal output is

$$P[\text{watt}] = 0.2 \times 10^{(-P[dB]/10)} \tag{4.3}$$

$$p[\text{dB}] = 10 \times \log(0.2 / P[\text{watt}]) \tag{4.4}$$

from which it can be once more seen that the dB scale is a logarithmic one: every 10 dB attenuation means an order-of-magnitude reduction in power. A good X-band bridge operates at power levels between 0 and −60 dB, i.e., between no attenuation and one million times attenuation. The electronic signal amplifier makes use of a discretely stepping potentiometer, which on all X-band spectrometers shows per decade the same "gain" values: 1.25, 1.6, 2, 2.5, 3.2, 4, 5, 6.3, 8, and 10. These are the rounded-off values of the operation $10^{0.n}$ for $n = 1$ through 10, which shows that the pot meter is a logarithmic device. Combining the logarithmic scale of the gain and the logarithmic scale of the power in dB leads to the following practical rule: the amplitude of a *non*-saturated EPR signal does not change if a reduction in power by 2 dB is compensated by an increase in gain by one step (or $2n$ dB compensated by n steps). This rule is used in the construction of the normalized power plot as in Figure 4.3C to determine P_M: the power is increased in steps of $2n$ dB, while the gain is concomitantly decreased by n steps until the measured EPR amplitude starts to decrease.

When the sample temperature is lowered below T_M, then for each temperature a new power plot has to be constructed. At lower T the EPR signal amplitude increases due to a more favorable Boltzmann distribution, however, this effect will be compensated by the fact that the maximal P value for which no saturation occurs will be less than P_M. At ever-lower T values, the maximally allowed P value moves towards the −60 dB limit, and eventually it becomes impossible to tune the spectrometer bridge. In fact, for a typical biological sample of 200 µl in a 4 mm inner diameter cylindrical quartz tube, spectrometer tuning and operation at powers below −40 dB is no sinecure, and the spectra can be easily deformed by the impossibility of avoiding mixing of an unwanted dispersion signal into the absorption signal. This problem is especially acute for the spectra of slowly relaxing $S = 1/2$ systems such as Cu^{2+} at low temperature ($T < 20$ K), which, however, can frequently be found in the literature, although such spectra do not usually sharpen up any further below circa 40–60 K. The bottom line of the above is that analysis of spectral shape and spin quantitation be preferably done close to T_M.

There may, however, be specific reasons to study a signal over an extended temperature range. For one, a linear increase in EPR amplitude with the inverse of the temperature (Curie's law) is proof that a spin system is a two-level system, i.e., an $S = 1/2$ or an effective $S = 1/2$ system. More importantly, in complex multicenter metalloproteins, overlapping spectra may be deconvoluted by virtue of their T_M value being different: if two centers, a and b, have $T_M^a < T_M^b$ then at T_M^b the spectrum of center a is broadened and that of center b is not. It is once more emphasized that these types of studies require determination of (anisotropic) saturation behavior at all relevant temperatures.

4.2 HOMOGENEOUS VERSUS INHOMOGENEOUS LINES

The Heisenberg uncertainty ΔE in the value of the energy of the excited state relative to the ground state (Equation 4.2) leads to a statistical distribution of the difference in energy between the ground state and the excited state and, therefore (cf. Equation 2.1), to a distribution in the frequency, and its associated wavelength, required for resonance. Since the magnitude of the uncertainty ΔE is a function of the sample temperature T, also the distribution in resonance frequency depends on T: the higher T, the wider the distribution; at $T = 0$ the distribution has its minimum width (not equal to zero!), eventually defined by the inequality in Equation 4.2. In all forms of spectroscopy this is observed as an intrinsic line shape with a width $\Gamma > 0$.

Mathematically, this line shape is described by the Lorentz distribution

$$F_{Lorentz}(\upsilon) = A_\upsilon \frac{1}{1+(\upsilon-\upsilon_0)^2/\Gamma_\upsilon^2} \tag{4.5}$$

in which v is the resonance frequency distributed around an average value v_0, A_v is the amplitude of the distribution at the central value v_0, and $2\Gamma_v$ is the full width at half maximum (FWHM). The "linewidth" is commonly reported at the half width at half maximum (HWHM), i.e., Γ_v, and it is the reciprocal of the relaxation time τ:

$$\Gamma_{Lorentz} \propto 1/\tau \tag{4.6}$$

The "intrinsic" in "intrinsic line shape" would seem to suggest that there is nothing much we can do about this nature-given distribution, but this is not really true. Not only can we tune its width by adjusting the temperature, but Equation 4.6 tells us that we can also influence Γ at fixed T by changing the relaxation rate τ. The latter is, however, much more difficult than a simple change of temperature: it requires a change of the coupling of the spin system with its surroundings, i.e., a change in molecular properties. In biological terms, any conformational change of a protein containing a paramagnet would change the relaxation rate and thus the linewidth of the Lorentzian distribution. There are many ways to induce a protein conformational change, but an experimentally easy one is to change the composition of the (frozen) medium by the addition of, e.g., 10% glycerol or 1–2 M urea to the solvent.

We have seen in Chapter 2 that in EPR spectroscopy one usually varies the magnetic field instead of the frequency, because the use of a mechanically rigid microwave resonator dictates the frequency to be constant. For this reason, the Lorentzian distribution in Equation 4.5 is frequently rewritten as a distribution in resonance fields as

$$F_{Lorentz}(b) = A \frac{1}{1+(b-b_0)^2/\Gamma^2}$$

$$\tag{4.7}$$

in which b is a variable field position around the central field position b_0. The linewidth, Γ, is now in field units (gauss). Formally, Equation 4.7 is not correct because the distribution is in E, therefore in v, therefore in $1/b$ (cf. Equations 2.1 and 2.3), but we will ignore the small error for the time being, and we will return to the subject of reciprocal distributions in Chapter 9 when discussing g-strain.

The Lorentzian line shape is also frequently called the *homogeneous* line shape to indicate that the distribution applies equally well to all molecules of a homogeneous sample, i.e., a sample of sufficient size (not just a few molecules) in which all molecules are chemically identical and indiscernible. The concept of chemical homogeneity is an idealization of reality, and it provides, as it turns out, an insufficient description of our real samples of, e.g., coordination complexes and, especially, metalloproteins. We know from folding studies and from structural NMR and x-ray studies that samples of proteins come with a distribution of conformations, which means that, although each molecule in the sample has the same chemical formula, the relative 3-D coordinates of a given atom, with respect to a reference atom in each molecule, are not exactly the same. For EPR this means that the paramagnet in each molecule has a slightly different structural surrounding and thus a slightly different g-value. This structural inhomogeneity is reflected in the spectroscopy in the form of an *inhomogeneous* line shape in addition to the homogeneous or Lorentzian shape. In fact, the "in addition" is not a proper expression because homogeneous broadening due to the Heisenberg uncertainty principle (physics) and inhomogeneous broadening due to structural inhomogeneity (chemistry) are independent phenomena, therefore they should not be added but be convoluted: to each position in the inhomogeneous distribution we should add a homogeneous distribution.

Conceptually, we are now facing a huge analysis problem: In order to be able to derive the exact form of an inhomogeneous distribution we would need detailed (and very complex) information on the distribution of structure (conformation) around the paramagnet plus a way to translate this data into a distribution of g-values. In practice, no attempt has ever been made yet in bioEPR to quantitatively derive such a distribution. Fortunately, we very frequently find experimentally that the EPR line shape at low temperature (i.e., when the contribution from homogeneous broadening is small) is very well reproduced by what is known as the Gaussian distribution (also called the normal distribution or the Laplace–Gaussian distribution). The use of the normal distribution as a model can be theoretically justified by assuming that many small, independent effects are additively contributing to an observation. Think, for example, of observing the g-value of a transition ion in a protein whose conformation is the net result of many small variations in bond lengths and angles between the thousands of atoms that make up the protein.

In terms of a resonance field b, the Gaussian distribution is

$$F_{Gauss}(b) = A \exp[-\ln 2(b - b_0)^2 / \Gamma^2] \tag{4.8}$$

in which the symbols have the same meaning as in the Lorentzian distribution in Equation 4.7, except that the linewidth Γ is not a reciprocal relaxation time, but is related to the Gaussian standard deviation, σ, by

$$\Gamma_{Gauss} = \sigma \sqrt{2 \ln 2} \tag{4.9}$$

Since in EPR we usually observe first-derivative spectra as a consequence of phase-sensitive detection (see 2.7) it is relevant to note that the first derivatives of the two distributions are features with a positive and a negative peak. The peak-to-peak separation Δ_{pp} in field units for the two distributions is

$$\Delta_{pp(Lorentz)} = 2\Gamma / \sqrt{3} \simeq 1.15\Gamma \qquad (4.10)$$

$$\Delta_{pp(Gauss)} = 2\Gamma / \sqrt{2\ln 2} \simeq 1.70\Gamma \qquad (4.11)$$

i.e., the derivative feature of a Lorentzian line, is sharper than the corresponding feature of a Gaussian line for the same linewidth. The two distributions and their derivatives on a field scale are compared in Figure 4.4.

In the practice of solid-state bioEPR, a Lorentzian line shape will be observed at relatively high temperatures and its width as a function of temperature can be used to deduce relaxation rates, while a Gaussian line will be observed at relatively low temperatures and its linewidth contains information on the distributed nature of the system. What exactly is high and low temperature, of course, depends on the system: for the example of low-spin cytochrome a in Figure 4.2, a Lorentzian line will be observed at $T \approx 80°C$, and a Gaussian line will be found at $T \approx 20°C$, while at $T \approx 50°C$ a "mixture" (a convolution) of the two distributions will be detected.

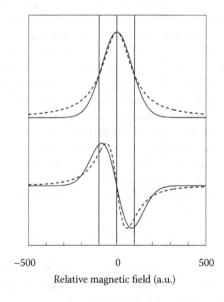

−500   0   500

Relative magnetic field (a.u.)

FIGURE 4.4 Line shapes. Lorentzian (broken lines) and Gaussian (solid lines) line shapes and their first derivatives are given. The outermost vertical lines delimit full width at half height (FWHH) of the absorption lines.

4.3 SPIN MULTIPLICITY AND ITS PRACTICAL IMPLICATIONS

A paramagnet with a single unpaired electron has a spin $S = 1/2$. However, many transition ion complexes and metalloproteins can, and do, have more than one unpaired electron. For example, the Fe(III) ion in ferrihemoglobin has five of them and its spin is equal to five times $1/2$ or $S = 5/2$. Usually, both the detection and the interpretation of the EPR of these high-spin systems is more complicated than that of $S = 1/2$ systems. We cannot afford to ignore these complications because roughly half of all paramagnetic metalloproteins are high-spin. In the paramagnetism of radicals high-spin configurations are rather less common, but we have already come across the inorganic biradical (i.e., two unpaired electrons) of triplet oxygen in solid air (see Chapter 3, Section 3.3). The system is called a triplet because it has three spin energy levels instead of the two levels of an $S = 1/2$ system.

A system with n unpaired electrons has a spin equal to $S = n/2$. Such a system has a spin multiplicity:

$$M_S = 2S + 1 \tag{4.12}$$

and this value is equal to the number of spin energy levels. All the spin levels together are called the *spin multiplet*. Table 4.1 lists a few simple examples together with their trivial names reminiscent of multifetal pregnancies or musical notes of unusual duration. An essential difference between $S = 1/2$ systems and high-spin or $S \geq 1$ systems is that the latter are subject to an extra magnetic interaction namely between the individual unpaired electrons. The electronic Zeeman interaction requires an external magnetic field B, absent when the spectrometer's magnet is switched off, but the mutual interaction between unpaired electrons is always present and is independent of any external field. Therefore, it is usually called the zero-field interaction. In biological transition ion complexes this zero-field interaction is usually (some Mn^{2+} complexes are the exception) stronger, and usually quite significantly so, than the Zeeman interaction produced by an X-band spectrometer.

This inequality of interactions leads to a phenomenon whose quantum-mechanical nature will be addressed later (Chapter 7), and whose practical appearance is given here simply by description. Let us first make the distinction between half-integer

TABLE 4.1
Spin multiplicities of selected transition ions

Ion	Spin	Multiplicity	Name
Cu^{2+}	1/2	2	Doublet
Ni^{2+}	1	3	Triplet
Co^{2+}	3/2	4	Quadruplet
Mn^{3+}	2	5	Quintuplet
Mn^{2+}	5/2	6	Hextuplet
Tb^{3+}	3	7	Heptuplet
Gd^{3+}	7/2	8	Octuplet

systems, i.e., systems with $S = n/2$ (3/2, 5/2, etc.), and integer systems, i.e., with $S = n(1, 2,$ etc.). This will turn out to be a very important distinction, indeed, because the EPR of half-integer systems is very different from that of integer systems. In zero field, the sublevels of a half-integer spin multiplet group in pairs, called *Kramer pairs*, and these pairs are separated by energy spacings significantly greater than the X-band microwave energy $h\nu$. These spacings are also called *zero-field splittings*, or ZFS. There is only one way to lift the pairing (also called *degeneracy*) of Kramer's levels: application of an external magnetic field. The resulting situation is schematically represented in the left-hand side of Figure 4.5 for an $S = 5/2$ system. The external field-induced splitting allows for the occurrence of EPR transitions within each (split) pair of levels; these are also known as *intradoublet transitions*. Since the number of Kramer's pairs is equal to

$$P_S = \left(S + \frac{1}{2}\right)\Big/2 \tag{4.13}$$

the number of possible intradoublet transitions is also equal to P_S; for S = 5/2 three such transitions are possible. Since the zero-field splittings between the Kramer's pairs are much greater than $h\nu$, very strong external magnetic fields are required to bring two sublevels of two different pairs sufficiently close together to allow for interdoublet transitions. The required field strengths are typically well beyond the maximum field of X-band spectrometers.

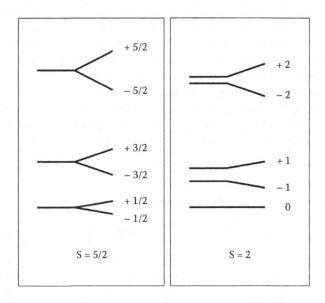

FIGURE 4.5 Examples of spin multiplets. Schematic outlines are given for the spin multiplets of S = 5/2 and S = 2 in zero field and in an external magnetic field of increasing strength. The doublets of a half-integer spin system are degenerate in zero field; those of an integer spin system are usually split even in zero field.

The labeling of the individual sublevels can easily give rise to confusion. By far the most common labeling is with the quantum number m_S, which runs in integer steps over the range

$$m_S = \{S, S-1, \ldots, -S\} \tag{4.14}$$

and in our example of $S = 5/2$ this means that we have $m_S = \{5/2, 3/2, 1/2, -1/2, -3/2, -5/2\}$ with the grouping in Kramer's pairs as indicated in Figure 4.5. Now a central theorem of EPR theory states that in a standard spectrometer, in which the z-component of the microwave magnetic field is perpendicular ($\perp$) to the z-axis of the external static field (as in Figure 2.4), transitions are allowed only between sublevels that differ by one unit in m_S:

$$\left| \Delta m_S \right|_\perp = 1 \tag{4.15}$$

but this so-called selection rule would imply that the only allowed transition would be the one within the $m_S = \pm 1/2$ doublet, and that all the other intradoublet transitions are forbidden, in other words that they have zero transition probability and thus zero intensity. This conclusion is falsified by a huge number of experimental observations (several will be illustrated in later chapters), and this implies that something is seriously wrong with our labeling of the sublevels using Equation 4.14. The explanation is that the used labeling is only rigorously applicable at infinitely high magnetic field strengths, and only approximately correct at relatively high field values, i.e., values typically beyond the maximum of X-band spectrometers. In the language of quantum mechanics, the m_S-values are "not good quantum numbers" at the B-values that we use in practice. The spin energy sublevels should be labeled with "linear combinations" of these quantum numbers. For example, the level that we label with $m_S = 3/2$ in Figure 4.5 should really be written as a linear combination of $+3/2$, $+1/2$, and $-1/2$, i.e., the level also has "some" $+1/2$ and "some" $-1/2$ character. Since a similar argument applies to the $m_S = -3/2$ level, it becomes possible to couple the two levels (i.e., to make an allowed EPR transition between them) by means of the selection rule in Equation 4.15. A practical problem is that we can only determine exactly how much the "some" character is by means of a full-blown quantum-mechanical analysis of the spectra of a particular system (we learn how to do this in Part 2). This requirement makes the correct labeling with linear combinations of m_S highly impractical, so we all stick to the high-field limit labeling of simple m_S values. In fact, some of us even allow ourselves the liberty of colloquially addressing the "$m_S = \pm 3/2$" intradoublet EPR spectrum as a "$|\Delta m_S| = 3$ transition," while being perfectly well aware that this is incorrect physics and a violation of the central selection rule in Equation 4.15.

Now let us turn to the spin energy sublevels of integer spin systems with $S = n$. Their spin multiplets differ from those of half-integer spins in two respects. Firstly, according to Equation 4.12, the spin multiplicity is an odd number, and so pairing up the individual sublevels will always leave one level ($m_S = 0$ in high-field notation) standing alone as illustrated in the right-hand of Figure 4.5. Moreover, the level pairs of integer spin systems are *non*-Kramer's doublets, which means that to lift their degeneracy does not necessarily require the presence of an external magnetic field. In fact, the zero-field interaction between the unpaired electrons is sufficient to create (relatively small) splittings between the two $m_S = \pm n$ levels (high-field notation!) of a non-Kramer's pair. The resulting mixing of levels makes them susceptible for an altogether different type of EPR transition, namely one caused by a microwave with its magnetic z-component *parallel* (∥) to the z-axis of the external magnetic field and with the selection rule

$$\left| \Delta m_S \right|_\parallel = 0 \qquad\qquad (4.16)$$

This type of spectroscopy requires a resonator that is different from the standard ones that we have met in Chapter 2. A rectangular cavity that is wider in the *b*-dimension (cf. Figure 2.5) is used to allow switching from operation in the TE_{102} perpendicular mode to the TE_{012} parallel mode. Such a resonator is called a bimodal cavity (see the pictures in Figure 4.6).

We began this chapter with a discussion of the two main external parameters to be adjusted during an EPR experiment, namely, power P and temperature T. We can now complete the set of key parameters by adding an intrinsic one: the spin S. EPR of systems with $S = 1/2$ is relatively easy not only theoretically but also experimentally, as it requires relatively low power levels and/or high temperatures. Half-integer spin systems with $S = n/2$ are more difficult, theoretically because they exhibit spin multiplets with transitions between each Kramer's pair, and experimentally because higher powers and lower temperatures are usually called for. Integer spin systems with $S = n$ are the ultimate challenge, theoretically because they have spin multiplets

FIGURE 4.6 Perpendicular versus bimodal cavities. The left hand set is from Bruker (ER 4102 ST and ER 4116 DM) and the right hand set is from Varian (E-231 and E-236); the latter look slightly more battered after three decades of service. Within each set the left hand resonator is the regular one (perpendicular mode) and the right hand resonator is the bimodal one (perpendicular and parallel mode). Note the increase in size for the bimodal cavities in the b-direction (defined in Figure 2.5).

of non-Kramer's pairs with complex zero-field splittings and transitions following an unusual selection rule, and experimentally because they require parallel-mode equipment and the application of high power levels and low temperatures. This classification in degrees of complexity is summarized in Table 4.2, together with a number of example systems.

TABLE 4.2

Classes of metalloproteins. Transition ion prosthetic groups in proteins are classified on the basis of their system spin and spin-lattice relaxation rate.

	$S = 0$	$S = 1/2$	$S = n/2$	$S = n$
Spin levels	One	One pair	$(n+1)/2$ pairs	$2n+1$ levels
Relaxation rate (T_1^{-1})	—	Slow	Faster	Fast
Onset of T-broadening (T_m)	—	77–RT	8–30	5–30
Cryo-coolant	—	Nitrogen	Helium	Helium
Examples (mononuclear)				
Manganese			Mn^{2+}, Mn^{4+}	Mn^{3+}
Iron	$lsFe^{2+}$	$lsFe^{3+}$	$hsFe^{3+}$	$hsFe^{2+}$, Fe^{4+}
Cobalt	Co^{3+}	$lsCo^{2+}$	$hsCo^{2+}$	
Nickel	$spNi^{2+}$	Ni^{1+}, Ni^{3+}		$hsNi^{2+}$
Copper	Cu^{1+}	Cu^{2+}		
Molybdenum	Mo^{6+}	Mo^{5+}		
Tungsten	W^{6+}	W^{5+}		
Examples (Fe/S clusters)				
Dinuclear (binuclear)	$[2Fe-2S]^{2+}$	$[2Fe-2S]^{1+}$		
Trinuclear		$[3Fe-4S]^{1+}$		$[3Fe-4S]^{0}$
Cubane	$[4Fe-4S]^{2+}$	$[4Fe-4S]^{1+}$	$[4Fe-4S]^{1+}$	
HiPIP[a]		$[4Fe-4S]^{3+}$		

[a] *Note:* A HiPIP stands for high potential iron sulfur protein, a trivial name to indicate that the reduction potential of this type of small ET proteins is relatively high: $+0.1 < E_{m,7} < +0.5$ volts. Formally, from an electrochemical viewpoint, the name is incorrect, because the E_m value applies to the $[4Fe-4S]^{(3+;2+)}$ transition, which is observed in aqueous solution of the protein, because it is much lower in value than that of cubanes in ferredoxins.

5 Resonance Condition

EPR spectra can reflect many different magnetic interactions, however, these all belong to a small set of basic forms resulting from pairwise interactions between magnets of three types: electron spins, nuclear spins, and laboratory magnets. EPR is a quantum-mechanical phenomenon; an efficient mathematical formalism known as the *spin Hamiltonian*, has been developed to describe the phenomenon and its resulting spectra. It is, however, not impossible to be a bioEPR spectroscopist (i.e., to apply EPR spectroscopy and to interpret the spectra in biochemical terms) without significant proficiency in quantum mechanics. In this chapter a set of resonance conditions is presented without explicit reference to the spin Hamiltonian; in other words, the expressions to describe EPR spectra are simply given, and their explanation and justification is deferred to Part 2 of this book. The given set is sufficient to describe, quantify, and (bio)chemically interpret most spectra.

5.1 MAIN PLAYERS IN EPR THEORY: B, S, AND I

We have seen in Chapter 2 that the electronic Zeeman term, the interaction between unpaired electrons in molecules and an external magnetic field, is the basis of EPR, but we have also discussed in Chapter 4 the fact that if a system has more than one unpaired electron, their spins can mutually interact even in the absence of an external field, and we have alluded to the fact that this zero-field interaction affords EPR spectra that are quite different from those caused by the Zeeman term alone. Let us now broaden our view to include many more possible interactions, but at the same time let us be systematic and realize that this plethora of possibilities is eventually reducible to five basic types only, two of which are usually so weak that they can be ignored.

To this goal we need to introduce, in addition to the external magnetic field B and the unpaired electron spin S as a magnet, another type of submolecular magnet: the nucleus with its nuclear spin I. It is, of course, well known from NMR that nuclei like ^1H, ^{13}C, or ^{15}N are magnetic, but in EPR, the magnetism of transition ion nuclei is at least as important as that of the light elements. Furthermore, high-resolution NMR is usually restricted to spin one-half nuclei ($I = 1/2$) but EPR does not make this discrimination: nuclei with $I > 1/2$ are equally welcome, because they are only slightly more difficult to deal with, both experimentally and theoretically, than $I = 1/2$ nuclei. For this reason, in EPR, contrast to NMR, ^{14}N ($I = 1$) is not usually replaced with ^{15}N ($I = 1/2$). And the very high nuclear spin of some elements (e.g., $I = 7/2$ for cobalt) is seen as a useful extra spectroscopic handle.

By far the most important influence of a nuclear spin on the EPR spectrum is through the interaction between the electron spin S and the nuclear spin I. Usually, at X-band frequencies this interaction is weaker, by an order of magnitude or more, than the electronic Zeeman interaction, and so it introduces small changes in the EPR spectrum known as hyperfine structure. As a first orientation to these patterns, note that just like the electron spin S, also the nuclear spin I has a multiplicity:

$$M_I = 2I + 1 \tag{5.1}$$

which means that the nucleus has a number of M_I different preferred orientations, and associated nuclear energy levels, in an external magnetic field. Here, the "external" applies to the nucleus itself, so the unpaired electron is also an "external" field. This means that the electron can experience not just one, but M_I different types of nuclei, each causing its own shift in the EPR resonance line, which consequently splits into M_I lines each with an (integrated) intensity equal to $1/M_I$ compared to that of an unsplit line. This is illustrated in Figure 5.1 for ^{51}V ($I = 7/2$). Note that the splitting between the eight lines is not constant; this is a "second order" effect that will be explained in Section 5.4. In Table 5.1, biologically relevant transition metal ions are listed with their nuclear spin; similarly, Table 5.2 gives the nuclear spin of biologically relevant ligand atoms.

Chemical bonds can have covalent character, and EPR spectroscopy is an excellent tool to study covalency: An unpaired electron can be delocalized over several atoms of a molecular structure, and so its spin S can interact with the nuclear spins I_i of all these atoms. These interactions are independent and thus afford additive hyperfine patterns. An unpaired electron on a Cu^{2+} ion ($S = 1/2$) experiences an $I = 3/2$ from the copper nucleus resulting in a fourfold split of the EPR resonance. If the Cu is coordinated by a

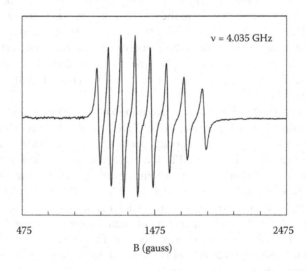

$\nu = 4.035\ \mathrm{GHz}$

475   1475   2475

B (gauss)

FIGURE 5.1 Isotropic hyperfine pattern for ^{51}V^{IV} in S-band. The spectrum is from VOSO$_4$ in aqueous solution. Use of the low frequency enhances the second-order effect of unequal splitting between the eight hyperfine lines.

TABLE 5.1
Biological metal transition ions and their nuclear spin

Metal	Isotope	Spin (abundance)	EPR lines
V	51	7/2	8
Mn	55	5/2	6
Fe	54, 56, <u>57</u>, 58	0 + 1/2(2%)	1 + 2(1%)
Co	59	7/2	8
Ni	58, 60, <u>61</u>, 62, 64	0 + 3/2(1%)	1 + 4(0.25%)
Cu	63, 65	3/2	4
Mo	92, 94, <u>95</u>, 96, <u>97</u>, 98, 100	0 + 5/2(25%)	1 + 6(4%)
W	180, 182, <u>183</u>, 184, 186	0 + 1/2(14%)	1 + 2(7%)

Note: Underlined isotopes have a nuclear spin $I \neq 0$.

nitrogen atom ($I = 1$) then each of the four lines will in turn be split (to a lesser extent) into three lines. If another ligand happens to be OH$^-$, then each of the 12 lines will be split (to an even lesser extent) into two by the proton ($I = 1/2$) resulting in 24 lines. Although these 24 lines need not (and usually will not) all be resolved, the point to make here is that chemically rather distinct phenomena such as the so-called central hyperfine interaction of a copper electron with its nucleus and the so-called superhyperfine interaction of that same copper electron with a ligand nucleus, from an EPR point of view, are of the same form. They are both interactions between an electron spin S and a nuclear spin I. This means that only a limited number of different types of terms are required to write out any resonance condition to describe EPR spectra.

In essence, one has to deal only with the above mentioned three types of magnets: B (manmade), S (electron spins), and I (nuclear spins). To avoid having to detail their specific forms at this time, let us indicate a pairwise interaction between these

TABLE 5.2
Biological ligand atoms and their nuclear spin

Ligand	Isotope	Spin (abundance)	EPR lines
H	1, 2	1/2; 1 (0.015%)	2 (& 3)
C	12, 13	0; 1/2 (1.1%)	1 (& 2)
N	14, 15	1; 1/2 (0.4%)	3 (& 2)
O	16, <u>17</u>, 18	0; 5/2 (0.04%)	1 (& 6)
F	19	1/2	2
P	31	1/2	2
S	32, <u>33</u>, 34	0; 3/2 (0.8%)	1 (& 4)
Cl	35, 37	3/2	4
As	75	3/2	4
Se	76, <u>77</u>, 78, 80, 82	0; 1/2 (7.6%)	1 (& 4)
Br	79, 81	3/2	4
I	127	5/2	6

Note: Underlined isotopes have a nuclear spin $I \neq 0$.

magnets with an asterisk, e.g., $S*B$ for the electronic Zeeman interaction. We then have six possible combinations, one of which, namely $B*B$, can be ignored as being trivial: as (bio)chemists we are not interested in the interaction between two different physical magnets. We are left with the five combinations $S*B$, $I*B$, $S*I$, $S*S$, and $I*I$. Some of their manifestations are given in Table 5.3. Two of these interactions depend on the external magnetic field, namely the electronic Zeeman interaction $S*B$ and the nuclear Zeeman interaction $I*B$. The latter is, of course, the basis for NMR spectroscopy, but since the nucleus is a much weaker magnet than the electron, we can safely ignore the contribution of $I*B$ to X-band bioEPR spectra except perhaps for octahedral Mn(II) complexes. Also, in very many cases, the relatively weak so-called quadrupole interaction $I*I$ within a magnetic nucleus does not significantly contribute to the cw EPR spectrum, and we can usually ignore it except perhaps for the rare cases in which arsenate (^{75}As has $I = 3/2$) has been used as a nonnatural ligand to replace natural oxoanions.

Similar to the $S*I$ hyperfine-interaction expressions being equally useful to describe central hyperfine and superhyperfine interactions, also a single set of $S*S$ zero field interaction expressions can be used to describe (bio)chemically rather different forms of interactions, for example, between electrons on a high-spin metal, between electrons of different paramagnetic (not necessarily high spin) prosthetic groups in a complex enzyme, and even between centers in different molecules such as dimeric metalloproteins (Figure 5.2). In chemical terms the obtained electronic information can thus be local (on a coordination complex), global (distance constraints between different centers of one molecule), or even supramolecular (between centers in different biomacromolecules).

Before we develop the resonance conditions for systems with hyperfine and with zero-field interactions, we return to the electronic Zeeman term $S*B$ as an example interaction to discuss a hitherto ignored complexity that is key to the usefulness of EPR spectroscopy in (bio)chemistry, namely anisotropy: the fact that all interactions

---

TABLE 5.3
Pairwise magnetic interactions in bioEPR spectroscopy (interactions given in typical decreasing order of strength at X-band frequencies)

Interaction	Phenomenon	Example
$S*S$	In high spin systems	$S = 5/2$ (Fe^{3+}), $S = 2$ (Fe^{2+})
	Exchange Intramolecular dipolar	Within metal clusters, for example,
	Intermolecular dipolar	$[2Fe-2S]^{1+}$ Between two $[4Fe-4S]^{1+}$ in ferredoxin Many model compounds (broadening)
$S*B$	Zeeman	Basic EPR pattern
	g-Strain	Inhomogeneous broadening
$S*I$	Metal hyperfine	^{57}Fe ($I = 1/2$), Cu ($I = 3/2$)
	Ligand hyperfine	^{1}H ($I = 1/2$), ^{14}N ($I = 1$)
$I*I$	Quadrupole interaction	Mn($I = 5/2$) and As($I = 3/2$) patterns
$S*I$	Nuclear Zeeman	In double-resonance spectra

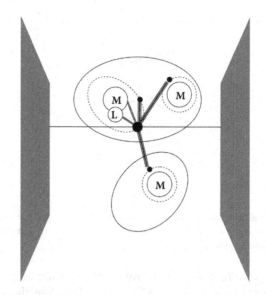

FIGURE 5.2 A schematic model of multiple $X*Y$ interactions. Black dots are unpaired electrons; the central, big black dot is the point of EPR observation. Straight lines are interactions: a single straight line symbolizes the electronic Zeeman interaction $S*B$; double lines represent central and ligand hyperfine interactions $S*I$; triple lines are zero-field interactions $S*S$ between electrons (i) around a single metal; (ii) at different centers within a molecule; and (iii) at centers in different molecules.

are spatially distributed or that the strength of the observed interaction depends on from what direction we look at a paramagnetic molecule. EPR spectra contain 3-D information on local symmetry and structure.

5.2 ANISOTROPY

Let us rewrite the resonance condition of an $S = 1/2$ system subject to the Zeeman interaction only as

$$hv = (g_e + \Delta g)\beta B \qquad (5.2)$$

to emphasize that the binding of an unpaired electron to a molecular system induces a shift away from the free-electron g-value $g_e = 2.00232$. Now consider an oxidized c-type cytochrome molecule with a low-spin ferric heme, whose tetrapyrrole structure we simplify to four nitrogens in a plane and at equal distance from the iron. The axial ligands are N (histidine) and S (methionine). This structure defines a Cartesian molecular axis system with the z-axis through the His-N and Met-S and with the x and y axis in the porphyrin plane (Figure 5.3). When this molecule is placed in an external dipolar magnetic field B with the field vector B along the molecular x axis, then the Δg in Equation 5.2 is largely determined by the "heteroaromatic" electrons of the porphyrin plane. When the magnet is realigned parallel to the molecular z axis (or equivalently, and perhaps more practically, when the molecule is rotated to have

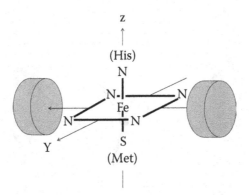

FIGURE 5.3 Axial anisotropy in an $S = 1/2$ system. A simplified representation is drawn of the porphyrin prosthetic group in low-spin ferricytochrome c in a magnet (Hagen 2006). (Reproduced by permission of The Royal Society of Chemistry.)

its z axis parallel with B), then Δg will have a different value as it is determined by the electronic structure of the His and Met ligands. Therefore, the magnitude of the experimental field value at resonance B is different for the two orientations. When the field vector is turned from x to z, its value changes smoothly between two extremes from B_x to B_z. On the other hand, when the vector is changed from x to y, nothing happens because the simplified porphyrin plane structure is symmetric, and $B_x = B_y$.

An X-band sample of 200 μL volume and 1 μM concentration contains circa $6 \times 10^{23} \times 200 \times 10^{-6} \times 1 \times 10^{-6} = 120{,}000{,}000{,}000{,}000$ paramagnetic molecules. In a sample of realistic size consisting of randomly oriented molecules, there are many molecules with the B vector anywhere in their xy plane, but there will be relatively few molecules with B parallel to the unique z axis. This notion provides a qualitative explanation of the intensity distribution in Figure 5.4, trace a. A real EPR spectrum is obtained when the stick spectrum is convoluted with a line shape with finite linewidth (trace b) and when the resulting EPR absorption pattern is differentiated with respect to the magnetic field, trace c. Note that the trace c in Figure 5.4 is easily misinterpreted to suggest that the EPR spectrum has two "peaks," whereas it actually has two features (or turning points) that correspond to points of (close to) zero slope in the EPR absorption spectrum, which also has intensity anywhere between the turning points.

When one of the Fe-coordinating Ns of the porphyrin is made inequivalent to the others, for example, by pulling on it, or by putting a protein structure around the cofactor, then the molecular x axis and y axis become inequivalent, and the axial EPR spectrum turns into the rhombic spectrum in trace d with derivative trace e (see also Table 5.4). There are now three features in the spectrum: a peak, a zero crossing, and a negative peak, and their field positions closely (exactly for zero linewidth) correspond to those of the g-values, g_x, g_y, and g_z. Finally, in trace f of Figure 5.4, which is the experimental X-band spectrum of cytochrome c, it can be seen that not only the g-value (peak position) but also the linewidth is frequently found to be anisotropic. This extra complication will be discussed extensively in Chapter 9.

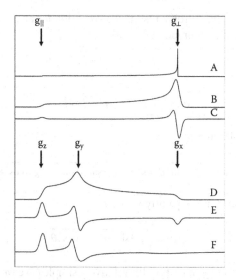

FIGURE 5.4 Anisotropy in absorption and derivative powder-type ERP spectra. (A) axial intensity pattern; (B) axial EPR absorption; (C) axial EPR derivative; (D) rhombic EPR absorption; (E) rhombic EPR derivative; (F) the spectrum of horse heart cytochrome c, a rhombic EPR derivative with anisotropic broadening (Hagen 2006). (Reproduced by permission of The Royal Society of Chemistry.)

Now let us become a bit more quantitative in our description of anisotropy by relating laboratory coordinates to molecular coordinates, more specifically by defining the dipolar magnetic field (a vector) of our electromagnet with respect to the coordinates of a molecule (and vice versa). The definition is through the two polar angles, θ and ϕ, where θ is the angle between the vector B and the molecular z axis, and ϕ is the angle between the projection of B onto the xy plane and the x-axis. An alternative, equivalent definition is in terms of the so-called direction cosines, l_i, that is, the cosines of the three angles between B and each of the molecular axes. The two definitions are related through

$$
\begin{aligned}
l_x &= \sin\theta\cos\phi \\
l_y &= \sin\theta\sin\phi \\
l_z &= \cos\theta
\end{aligned}
\tag{5.3}
$$

TABLE 5.4

Types of spectra based on symmetry

Symmetry	Characteristics	g-Values
Isotropic	$g_x = g_y = g_z$	1
Axial	$g_x = g_y \neq g_z$	2
Rhombic	$g_x \neq g_y \neq g_z$	3

Now we can define the anisotropic resonance condition for an $S = 1/2$ system subject to the electronic Zeeman interaction only as

$$B_{res} = hv / g(l_x, l_y, l_z)\beta \qquad (5.4)$$

in which

$$g(l_x, l_y, l_z) = \sqrt{l_x^2 g_x^2 + l_y^2 g_y^2 + l_z^2 g_z^2} \qquad (5.5)$$

or in terms of the polar angles

$$g(\theta, \phi) = \sqrt{g_x^2 \sin^2\theta\cos^2\phi + g_y^2 \sin^2\theta\sin^2\phi + g_z^2 \cos^2\theta} \qquad (5.6)$$

which, for axial spectra, is frequently written as

$$g_{ax}(\theta) = \sqrt{g_\perp^2 \sin^2\theta + g_\parallel^2 \cos^2\theta} \qquad (5.7)$$

in which $g_\perp \equiv g_x \equiv g_y$, and $g_\parallel \equiv g_z$. For this relatively simple case we have plotted θ and $g_{ax}(\theta)$ as a function of the resonance field B_{res} in Figure 5.5 to graphically illustrate the important point that the resonance field B_{res} is relatively insensitive to changes in orientation (here, in θ), that is, $\delta B_{res}/\delta\theta \approx 0$, for orientations of B near the molecular axes (here, $\theta = 0$ or $\pi/2$). This is a mandatory condition for clear so-called turning-point features to appear in the powder spectrum and specifically for the peaks and zero crossing of a rhombic spectrum to closely correspond to the three g-values, and for the peak and the point at two-thirds between top and bottom of the derivative feature of an axial spectrum to closely correspond to the two g-values. We will later come across more complicated cases for which this condition does not hold, which means that the g-values (and other EPR parameters) are not readily read from the spectrum but can only be obtained by spectral simulation.

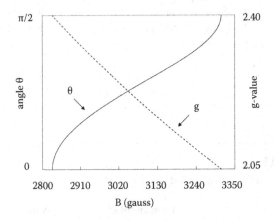

FIGURE 5.5 Angular dependency of axial g-value. The angle θ between B_0 and the molecular z-axis and the axial g-value are plotted versus the resonance field for a typical tetragonal Cu(II) site with $g_\parallel = 2.40$ and $g_\perp = 2.05$; $v = 9500$ MHz.

5.3 HYPERFINE INTERACTIONS

The spectrum given in Figure 5.1 as a first example of hyperfine structure is a single-line spectrum split by interaction with a nuclear spin. In these types of spectra, both the Zeeman term and the hyperfine term are isotropic (i.e., independent of orientation) because all anisotropy is averaged away by rapid tumbling of the compound in water. Biomacromolecules do not tumble sufficiently rapid at ambient temperatures for this complete averaging to occur. And molecules in frozen solutions do not tumble at all. Therefore, all magnetic interactions determining the EPR spectrum are generally aniso-tropic, and we need proper resonance-condition expressions to describe the spectra. Once again, we note that all expressions given in this chapter are, for the time being, axiomatic: they come out of the blue and their value is only verified because they turn out to be "good" in practice.

When the hyperfine interaction is much smaller than the Zeeman interaction ("much" means approximately two orders of magnitude or more), as is usually the case in X-band, then the resonance condition is

$$h\nu = g\beta B + A'm_I = g\beta(B) + Am_I \quad (5.8)$$

Be particularly careful with the units here: A' is an energy in units of cm^{-1}, whereas A is in magnetic-field units of gauss, that is, $A = A'/(g_e\beta)$. From now on we will always express hyperfine splittings in magnetic-field units, because this is what we read out from the field-swept spectrum. In most EPR textbooks hyperfine splittings are given in energy units and the symbol used is A, not A'. In some texts the splitting in field units has the symbol a_0 (Weil and Bolton 2007). We prefer to reserve the symbol a for the higher-order cubic zero-field splitting parameter (Chapter 8).

The selection rules are

$$\Delta m_S = 1 \text{ and } \Delta m_I = 0 \quad (5.9)$$

and since we have $(2I+1)$ nuclear levels, for $S = 1/2$ there are also $(2I+1)$ transitions. In terms of a resonance field we have

$$B_{res} = (h\nu / g\beta) - \sum_{m_I} Am_I \quad (5.10)$$

and note that for half-integer nuclear spins m_I is never equal to zero, which means that no peak will occur at the g-value (i.e., at a field corresponding to $B = h\nu/g\beta$). This is, for example, the case in the V(IV) spectrum in Figure 5.1 where eight lines are approximately equidistantly grouped around the vanadyl g-value of 1.96. Note that for a frequency of 4.0 GHz, this g-value corresponds to a field of circa 1460 gauss, while the hyperfine splitting between the peaks is $A = 111$ gauss, in other words $S*I$ and $S*B$ differ by only one order of magnitude. This implies the neces-sity of a second-order correction to Equation 5.10, which is discussed in Section 5.4. In the spectrum of Figure 5.1 this so-called "second-order effect" is observed as a nonequidistant splitting between the eight hyperfine lines.

In any metalloprotein, be it tumbling in water or fixed in a frozen solution, not only the Zeeman interaction but also the hyperfine interaction will be anisotropic, so the resonance field in Equation 5.10 becomes a function of molecular orientation in the external field (or alternatively of the orientation of B in the molecular axes system):

$$B_{res}(l_x,l_y,l_z) = \sum_{m_I} \{[h\nu / g(l_x,l_y,l_z)\beta] - A(l_x,l_y,l_z,g_x,g_y,g_z)m_I\} \tag{5.11}$$

in which $g(l_x,l_y,l_z)$ was already defined in Equation 5.5, and

$$A(l_i g_i) = g^{-2}\sqrt{l_x^2 g_x^4 A_x^2 + l_y^2 g_y^4 A_y^2 + l_z^2 g_z^4 A_z^2} \tag{5.12}$$

with g as defined in Equation 5.5, or written out in full for polar angles

$$A(\theta,\phi) = \frac{\sqrt{\sin^2\theta\cos^2\phi g_x^4 A_x^2 + \sin^2\theta\sin^2\phi g_y^4 A_y^2 + \cos^2\theta g_z^4 A_z^2}}{\sin^2\theta\cos^2\phi g_x^2 + \sin^2\theta\sin^2\phi g_y^2 + \cos^2\theta g_z^2} \tag{5.13}$$

which may look like a pretty hairy beast, but this expression will turn out in the next chapter to be quite trivial to handle for your PC.

Watch out: In many physics textbooks on EPR (e.g., Abragam and Bleaney 1970), you will not find Equation 5.12 (or its equivalent Equation 5.13) but a different expression, namely:

$$A(l_i g_i) = [g(l_i)]^{-1}\sqrt{l_x^2 g_x^2 A_x^2 + l_y^2 g_y^2 A_y^2 + l_z^2 g_z^2 A_z^2} \tag{5.14}$$

which is also written as

$$g^2 A^2 = l_x^2 g_x^2 A_x^2 + l_y^2 g_y^2 A_y^2 + l_z^2 g_z^2 A_z^2 \tag{5.15}$$

These expressions are valid only for frequency-swept spectra, and the As are in cm^{-1} units. And since we always sweep the field, while keeping the frequency fixed, the Equations 5.14 and 5.15 are practically useless.

Phenomenologically, the most apparent manifestation of anisotropy in the central hyperfine splitting is that it is frequently much better resolved in one direction, that is, along a particular molecular axis than along the other two directions. We have already seen an example of this pattern in the X-band spectrum of the hydrated Cu(II) ion in Figure 3.4. This is an axial spectrum with $g_{||} > g_\perp$, that is, the $g_{||}$-peak is on the low-field side of the spectrum. Actually, there is no peak at $g_{||}$ because the copper nucleus has a half-integer spin $I = 3/2$ and so we find four hyperfine lines around the $g_{||}$-value with $A_{||} = 130$ gauss. In the perpendicular direction around $g_\perp$ we find a single derivative feature: the hyperfine interaction in this direction (i.e., in the molecular xy plane) is not resolved. In other words, $A_\perp$ is significantly less than the linewidth $W_\perp$.

Another illustrative example is that of the Co(II) ion in vitamin B$_{12}$ (cyanocobalamin) given in Figure 5.6. The cobalt is coordinated by four nitrogen atoms from

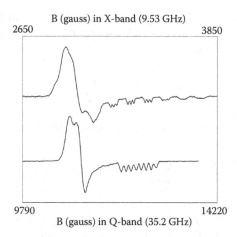

B (gauss) in X-band (9.53 GHz)

2650   3850

9790   14220

B (gauss) in Q-band (35.2 GHz)

FIGURE 5.6 X-band and Q-band EPR of Co(II) in cyanocobalamin. Vitamin B_{12} (cyano-cobalamin) in aqueous solution was reduced with a grain of $NaBH_4$. The approximate EPR spectral parameters are g_{xyz} = 2.31, 2.26, 2.02; A_z (Cu) = 114 gauss; A (N) = 19 gauss.

a tetrapyrrole system spanning an approximate plane like in the case of cytochrome c in Figure 5.3, and with an axial cyano N ligand. The system is slightly rhombic ($g_x \neq g_y$) as can be seen in the lower spectrum in Figure 5.6. The g_z-value is now on the high-field side and its negative peak is split by the cobalt nucleus (I = 7/2) in eight lines with A_z = 114 gauss. Again, A_x and A_y are less than the linewidth and thus not resolved. Note, however, that this spectrum was not recorded in X-band but at a four times higher frequency in Q-band (circa 35 GHz). The corresponding X-band spectrum is in the upper trace of the figure, and the two spectra combined show some typical effects of multifrequency EPR (here, the "multi" is equal to two). The $S*B$ interaction is dependent on the strength of the external magnetic field, the $S*I$ interaction is not. Decreasing the microwave frequency, as we do in Figure 5.6 from the bottom trace to the top trace, relatively decreases the importance of the Zeeman term with respect to the hyperfine term. From Q-band to X-band the hyperfine splitting A_z remains constant in field units, but the separation of the g-values in field units decreases. We observe this in the X-band spectrum as (1) a loss of g-value resolution, and (2) a partial overlap (or interference) of the A_z hyperfine pattern with the g_{xy} peak leading to an apparently rather complex powder pattern. On the other hand, the lowering of the frequency has also resulted in a decrease in linewidth, and an extra pattern of three lines starts to appear in some of the cobalt hyperfine lines. This is the superhyperfine (or ligand hyperfine) interaction of the unpaired d-electron of the cobalt with the axial cyano nitrogen ligand (I = 1). The bottom line is that playing around with the frequency changes the relative weight of the different interaction terms, and thus may help in resolving individual interactions and in interpreting the spectra. Do note, however, that this "playing around" is never trivial from a financial and organizational viewpoint because each extra frequency requires an extra spectrometer.

The example of nitrogen lines in the spectrum of cobalamin points to the necessity of also writing out resonance conditions for the presence of ligand hyperfine interaction. In general we have:

$$h\nu = g\beta B + Am_I + \sum_{ligand} A_L m_{I_L} \qquad (5.16)$$

in which the summation is over all possible ligands. In terms of the resonance field this is:

$$B_{res} = \sum_{ligand} \sum_{m_{I_L}} \sum_{m_I} [(h\nu / g\beta) - Am_I - A_L m_{I_L}] \qquad (5.17)$$

and the anisotropy is introduced as previously by writing g as in Equation 5.5 and writing the central hyperfine A and each of the ligand hyperfine A_L terms as in Equation 5.12. Analyzing such a system could be an involved enterprise where each summation in Equation 5.17 can be read as an order-of-magnitude increase in complexity, and one may well start to wonder whether the biochemical turnout is worth the effort. Fortunately, in practice superhyperfine splittings in metalloprotein EPR are often found to exhibit rather limited anisotropy, and, for example, acceptably good fits to experimental data are usually obtained by assuming isotropic interactions.

5.4 SECOND-ORDER EFFECTS

In introducing the resonance condition for a system with hyperfine interaction we made the restriction that it should be much smaller than the Zeeman interaction for the Equation 5.8 to be valid. This was an allusion to the fact that all resonance conditions given in the present chapter are analytical expressions derived using a technique called perturbation theory. This approach starts from a dominant interaction (here, the Zeeman term) and treats a weaker interaction (here, the hyperfine term) as a small perturbation. The result of such a treatment is a relatively simple expression for the case when the perturbation is very small, a significantly less simple expression when the perturbation is perhaps not so small, and expressions of rapidly increasing complexity when the perturbation becomes comparable in magnitude to the main interaction. We indicated that "very small" means that the perturbation is approximately two orders of magnitude less than the main interaction, which, for example, for an $S = 1/2$ system in X-band (9.5 GHz) with a g-value of circa 2 resonating at a field of circa 3400 gauss would mean a hyperfine splitting of the order of 34 gauss.

Some transition ions have central hyperfine splittings somewhat greater than this value, for example, for copper one typically finds A_z values in the range 30–200 gauss, and so in these systems the perturbation is "not so small," and one has to develop so-called second-order corrections to the analytical expression in Equation 5.12 or 5.13 that is valid only for "very small" perturbations. The second-order perturbation result (Hagen 1982a) for central hyperfine splitting is:

$$B_{res}^{(2nd)} = (B_{res} / 2) + (1 / 2)\sqrt{B_{res}^2 + K_1 + K_2 m_I^2} \qquad (5.18)$$

in which the first-order B_{res} is defined in Equation 5.11 (and 5.5 and 5.12), and

$$K_1 = (A_1^2 - K_0)I(I+1) \tag{5.19}$$

$$A_1 = g^{-3}A^{-1}\sqrt{l_x^2 g_x^6 A_x^4 + l_y^2 g_y^6 A_y^4 + l_z^2 g_z^6 A_z^4} \tag{5.20}$$

$$K_0 = g^{-2}(g_x^2 A_x^2 + g_y^2 A_y^2 + g_z^2 A_z^2) \tag{5.21}$$

$$K_2 = 2A^2 - A_1^2 - K_0 \tag{5.22}$$

Note that, just like for the first-order expression in Equation 5.12 also the second-order expression in Equation 5.18 applies to field-swept spectra, and a different expression found in EPR textbooks (Pake and Estle 1973) applies to frequency-swept spectra. The effect of including a second-order contribution to the central hyperfine splitting is illustrated in Figure 5.7 on the spectrum of a not uncommon contaminant of metalloprotein preparations: Cu(II) ion coordinated by nitrogens of tris-hydroxyethyl aminomethane or Tris buffer.

What would happen if we were to lower the microwave frequency from X-band to L-band (1 GHz). The Zeeman term for $g \approx 2.2$ (an average value for copper) would correspond to a field of circa 325 gauss at 1 GHz, and so the two interactions $S*B$ and $S*I$ would be of comparable magnitude. In such situations the perturbation expressions become extremely complicated and lose all practical significance.

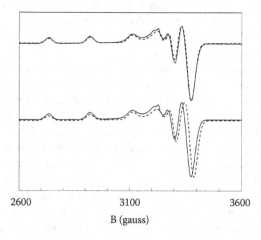

2600   3100   3600

B (gauss)

FIGURE 5.7 Second-order hyperfine shift in the X-band EPR of the Cu(II)-Tris complex. The thin solid line is the experimental spectrum of 1.5 mM CuSO$_4$ in 200 mM Tris-HCl buffer, pH 8.0 taken at ν = 9420 MHz and T = 61 K. Tris is tris-(hydroxymethyl)aminomethane or 2-amino-2-hydroxymethyl-1,3-propanediol. The broken lines are simulations using the parameters: $g_\perp$ = 2.047, $g_{||}$ = 2.228, $A_{||}$ = 185 gauss. In the lower trace the second-order correction has been omitted.

It then becomes unavoidable to analyze the spectra numerically using matrix diagonalization methods rooted in quantum mechanics. We can generalize this conclusion
as follows: whenever the main interaction determining an EPR spectrum becomes
comparable in magnitude to a second interaction (or to more interactions), then the
resonance expressions of this chapter are no longer applicable, and the analysis
requires a numerical approach, to be treated in Part 2 of this book. Fortunately, very
many of the X-band metalloprotein EPR spectra reported thus far do not fall into this
"difficult" category, although it is possible that the number of difficult cases thus far
reported is relatively small simply because it may be experimentally more demanding to collect this type of data.

5.5   LOW-SYMMETRY EFFECTS

We used direction cosines l_i in the expressions for anisotropy in the g-value and in
hyperfine A-values. These l_i's were defined with respect to a molecular Cartesian
axes system. How is this axes system actually defined? In the several examples of
tetrapyrrole ligands, above, an obvious assignment would be to draw—as we did in
Figure 5.3—two axes (x and y) either through the four in-plane nitrogen atoms or in
between these four atoms, and to take the z-axis as a line through the axial ligand(s).
This choice implies an idealized structure with the tetrapyrrole N's on the corners of
a perfect square and with axial ligands exactly on a line perpendicular to the center
of this square (for example, forming an elongated octahedron); however, a real situation is not expected to deviate too much from this picture. But what if there is no
tetrapyrrole ring and, for example, a transition metal ion is directly bound to a few
amino acid side groups with no clearly identifiable regular structure at all? It turns
out that even for these systems a unique molecular axes system exists, defining the
direction cosines required for our anisotropy expressions, but it is not obvious from
inspection of the molecular structure (for example, via a protein database file) how to
place this axes system in the molecule.

It is possible to determine the direction of the axes by single-crystal EPR, that is,
one must first crystallize the compound and then measure the g-value as a function
of orientation of the crystal in the magnetic field of the spectrometer. In practice,
this can be a very time consuming and difficult task, for example, because single
crystals of metalloproteins are usually too small to get a decent signal out of an
X-band spectrometer. From a biochemical viewpoint one could decide that the orientation of the molecular axis system is not a very exciting entity that we can easily
live without, and this is in fact the approach commonly taken. The spectra can still
be used as a fingerprint and for concentration determination (see below). There is,
however, one complication that we cannot ignore: in low-symmetry systems the axis
system that defines the anisotropy in the g-value need not necessarily be the same
axis system that defines, for example, the anisotropy of a central hyperfine system.
In other words, the l_i's and (θ,ϕ)-pair that we use in the g-value Equations 5.5 and 5.6
may well be completely different from a set of (l_x', l_y', l_z') and a pair (θ',ϕ') required to
describe the A-value in Equations 5.12 and 5.13. In EPR lingo this is usually referred
to as *tensor noncolinearity*. We will later see in Chapter 8 that mathematically this

is not a difficult problem to deal with, however, finding a unique interpretation of the resulting EPR spectra by simulation becomes a tedious task. For now, let us just list a number of spectral effects to be expected for low-symmetry systems with reference to the example of Cu(II) bound to the iron transport protein transferrin (the protein structure does not change by this metal substitution; see Smith et al. 1992), whose X-band spectrum is given in Figure 5.8. The peak at g_z is split into four copper hyperfine lines but the splitting A_z^* is indicated with an asterisk to indicate that its value is not necessarily a turning point in the $A(\theta',\phi')$ plot: there is an $A_{z'}$-value (note the little superscript prime at the subscript z) for B along the z'-axis ($\theta' = 0$) with $A_{z'} > A_z^*$. Also, at high field a negative peak is observed suggesting a rhombic spectrum, but simulations indicate that this interpretation is not correct, because in a real rhombic spectrum the peak should have higher (negative) amplitude than the negative lobe of the derivative feature on its low-field side. The extra peak is simply a consequence of low symmetry. This also holds for the extra asymmetric peaks observed in between the four A_z^* hyperfine lines, which might easily be mistaken as a sign of sample inhomogeneity.

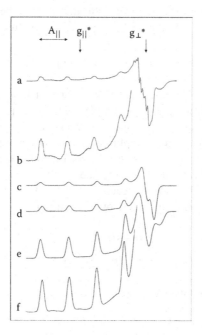

FIGURE 5.8 Complex hyperfine patterns due to axes noncolinearity in a low-symmetry prosthetic group. The X-band spectrum is from ^{65}Cu(II)-bicarbonate in human serum transferrin: (a,b) experimental spectrum; (c,e) simulation assuming axial symmetry; (d, f) simulation assuming triclinic symmetry with the A-axes rotated with respect to the g-axes over 15° about the g_z-axis and then 60° about the new y'-axis. Traces b, e, and f are 5× blow-ups of traces a, c, d, respectively (Hagen 2006). (Reproduced by permisson of The Royal Society of Chemistry.)

5.6   ZERO-FIELD INTERACTIONS

We have seen that the $S*S$ term can formally describe many different types of interaction, but perhaps the most common one is the intraelectron interaction between the unpaired electrons of a high-spin system. The perturbation-theory analysis of these systems is different from that previously shown for systems subject to hyperfine interaction, because we now usually find $S*S \gg S*B$ in X-band (i.e., the zero-field term dominates and the Zeeman term is the perturbation). In the previous chapter we indicated that this situation leads to Kramer's doublets for half-integer spins ($S = n/2$) and to non-Kramer's doublets for integer spins ($S = n$). Transitions are possible within these doublets (intradoublet) but not between the doublets (interdoublet) because the microwave energy $h\nu$ is much smaller than the zero-field splittings. Contrast to hyperfine spectra the resonance lines are not split, but they do, of course, occur at resonance fields different from those in the absence of zero-field interactions. This change is described by means of effective g-values or peak positions described by the condition

$$h\nu = g^{eff}\beta B$$

(5.23)

in which g^{eff} encompasses the real g-value plus the effect of the zero-field interaction. We could have also taken this approach for hyperfine structure by rewriting Equation 5.8 as

$$g^{eff} = (h\nu - Am_I)/\beta B$$

(5.24)

but this would have the disadvantage that we would have to report an effective g-value for each hyperfine line (for example, eight g_i^{eff}'s per direction for $I = 7/2$ nuclear spins) instead of one g-value and one hyperfine splitting. Since the resonance lines for $S*S \gg S*B$ systems are not split, we do not have that problem here.

Just like the g-value and A-values also the zero-field interaction parameter can be anisotropic and have three values D_x, D_y, and D_z. In contrast to g and A, however, the three D_i's are not independent because $D_x^2 + D_y^2 + D_z^2 = 0$, and so they can be reduced to two independent parameters by redefinition:

$$D = 3D_z/2$$

$$E = (D_x - D_y)/2$$

(5.25)

We can also define a rhombicity

$$\eta = E/D$$

(5.26)

which on theoretical ground turns out to be limited to the range $0 \leq \eta \leq 1/3$ (Troup and Hutton 1964). For axial symmetry $E = 0$ and so $\eta = 0$, and first-order perturbation theory gives expressions for the effective g-values, namely for the $m_S = \pm 1/2$ doublet

$$g_\perp^{eff} = (S + 1/2)g_\perp$$

(5.27)

$$g_\parallel^{eff} = g_\parallel$$

(5.28)

and for the other doublets

$$g_\perp^{\text{eff}} = 0 \tag{5.29}$$

$$g_\parallel^{\text{eff}} = (|\, m_S \,| + 1/2) g_\parallel \tag{5.30}$$

and since an effective g-value equal to zero corresponds (Equation 5.4) to an infinitely large resonance field, all spectra of interdoublet transitions other than the $m_S = \pm 1/2$ doublet are spread out over an infinite field range, which implies that their amplitude must be zero. In other words, only the transition within the $\pm 1/2$ doublet is observable. For example, assuming a real g-value close to the free-electron g_e-value (i.e., $g_\perp = g_\parallel = 2$) we would find for an $S = 3/2$ system a single transition with effective g-values $g_\parallel^{\text{eff}} = 2$ and $g_\perp^{\text{eff}} = 4$. And for $S = 5/2$ with $g = 2$ we would find $g_\parallel^{\text{eff}} = 2$ and $g_\perp^{\text{eff}} = 6$. Examples are the spectrum of the Fe(II)EDTA-NO complex ($S = 3/2$) and the spectrum of Fe(III)myoglobin ($S = 5/2$).

The axial effective g-values for real $g = 2$ are summarized in the scheme in Figure 5.9 for $S = 3/2$ through $S = 9/2$. This scheme is a useful aid to memory for the rapid recognition of many high-spin spectra not only in full axial symmetry. The introduction of a small rhombicity, $E \neq 0$, has the following qualitative effects. For the $\pm 1/2$ doublet the $g_\perp^{\text{eff}}$ will split into $g_x^{\text{eff}} < g_\perp^{\text{eff}}$ and $g_y^{\text{eff}} > g_\perp^{\text{eff}}$; the $g_\parallel^{\text{eff}}$ of the $\pm 1/2$

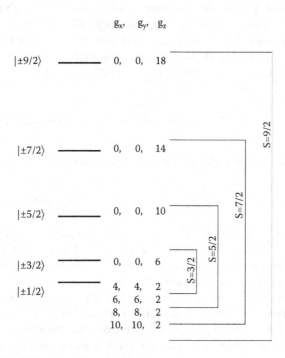

FIGURE 5.9 Effective g-values for half-integer spin systems in axial symmetry. The scheme gives the g-values for all transitions within the Kramer's doublets of $S = n/2$ systems assuming $g_{\text{real}} = 2.00$ and $S*S \gg S*B$.

doublet will become less than $g_\parallel$ with increasing E. For relatively small values of E the analytical expressions to first order become

$$g_x^{\text{eff}} = (S + 1/2)g_x[1 - (4E/D)]$$ (5.31)

$$g_y^{\text{eff}} = (S + 1/2)g_y[1 + (4E/D)]$$ (5.32)

$$g_z^{\text{eff}} = g_z$$ (5.33)

For example the X-band spectra of high-spin ferric hemes ($S = 5/2$; $g \approx 2$; $E/D < 0.1$) show a single transition described by

$$g_\perp^{\text{eff}} \approx 6 \pm 24E/D$$ (5.34)

$$g_\parallel^{\text{eff}} \approx 2$$ (5.35)

With increasing rhombicity the g_x^{eff} and g_y^{eff} of higher Kramer's doublets increasingly move away from zero and so the spectra of higher intradoublet transitions become spread over decreasing field ranges; that is, they gain intensity. Perhaps the most well known example of this behavior is the $g^{\text{eff}} = 30/7 \approx 4.3$ line from the $S = 5/2$ system Fe(III) in a rhombic environment with $E/D \approx 1/3$, which appears in the EPR spectrum of every sample of biological origin (the material is referred to as adventitious iron, dirty iron, crap iron, or even shit iron). This is a line of relatively strong intensity even for low Fe(III) concentrations, because for $E/D = 1/3$ the effective g-values of the $m_S = \pm 3/2$ doublet happen to all three coincide according to the expression (Aasa 1970)

$$g_x^{\text{eff}} = g_x[(15/7) - 60(102/2401)\{(1 - 3E/D)/(1 + E/D)\}^2]$$ (5.36)

$$g_y^{\text{eff}} = g_y[(15/7) - (60/49)(1 - 3E/D)/(1 + E/D)]$$ (5.37)

$$g_z^{\text{eff}} = g_z[(15/7) + (60/49)(1 - 3E/D)/(1 + E/D)]$$ (5.38)

This expression rapidly loses its validity with E/D values decreasing from full rhombicity. Similarly, when starting from axial symmetry for the $m_S = \pm 1/2$ doublet at increased rhombicity values the Equations 5.31–5.33 become increasingly inaccurate, and for $E/D > 0.1$ they are no longer valid.

In general, no simple, consistent set of analytical expressions for the resonance condition of all intradoublet transitions and all possible rhombicities can be derived with the perturbation theory for these systems. Therefore, the rather different approach is taken to numerically compute all effective g-values using quantum mechanics and matrix diagonalization techniques (Chapters 7–9) and to tabulate the results in the form of graphs of g^{eff}'s versus the rhombicity $\eta = E/D$. This is a useful approach because it turns out that if the zero-field interaction is sufficiently dominant over the Zeeman interaction ($S*S \gg S*B$), then the effective g-values become completely independent of the magnitude of D; they only depend on the ratio E/D and on the real g-values. For systems with half filled valence shell such as $3d^5$ Fe^{III} the real g-values are very close to the free electron value, and consequently, the EPR spectra become a function of a single parameter only: the rhombicity η. Figure 5.10 is a "rhombogram," that is, a plot of all effective g-values as a function of η, for $S = 5/2$. Interpretation of experimental spectra is done by moving a vertical ruler over the rhombogram until a fit to experimentally observed effective g-values is obtained. For example, the vertical line drawn for $\eta = 0.24$ corresponds to the spectrum of Fe^{III} in the enzyme superoxide dismutase (Figure 5.11) and is made up of three subspectra from three intradoublet transitions, whose relative intensities change with changing temperature.

The inequality $S*S \gg S*B$ is sometimes called the "weak-field limit" because the external magnetic field B of a typical X-band spectrometer is sufficiently weak for

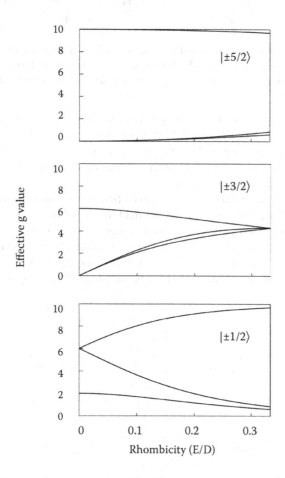

FIGURE 5.10 The rhombogram for $S = 5/2$. Effective g-values of the three intradoublet transitions have been plotted as a function of the rhombicity $\eta = E/D$ assuming $g_{real} = 2.00$ and $S*S \gg S*B$.

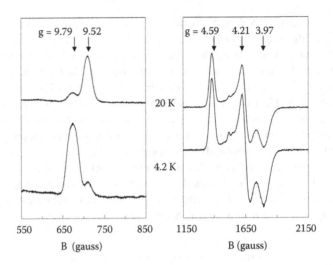

FIGURE 5.11 EPR of an $S = 5/2$ system with pronounced rhombicity. The X-band spectra ($v = 9.31$ GHz) are from high-spin Fe^{III} in *Escherichia coli* iron superoxide dismutase. The observed effective g-values correspond to $\eta = 0.24$ in the $S = 5/2$ rhombogram.

the inequality to be true. For some systems (for example, five-coordinate Mn^{II} in proteins; Fe^{III} in Al_2O_3) the zero-field interaction is not so strong and at X-band the two interactions are comparable in magnitude: $S*S \approx S*B$. One could call this "the intermediate-field" situation (cf. Chapter 12), and the spectra can only be analyzed with quantum-mechanical methods.

It is also possible that the zero-field interaction is much weaker than the Zeeman interaction: $S*S \ll S*B$, and this "strong-field limit" holds for six-coordinate Mn^{II}, which is not only biologically relevant as a site in some manganese proteins, but also because this is another very common contaminant of biological preparations; "adventitious manganese" or "dirty manganese" (see Figure 5.12). This system is in fact rather complex because manganese had $I = 5/2$ and the central hyperfine

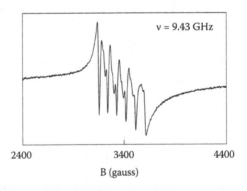

FIGURE 5.12 Manganese as a common contaminant in protein EPR. This X-band spectrum is characteristic for high-spin Mn(II) aspecifically bound to proteins.

interaction is of the same order as the zero-field interaction, that is, $S*B >> S*S \approx S*I$, and there is even a small contribution from the $I*B$ nuclear Zeeman term (and a very small $I*I$ contribution). The spectrum is well understood from analysis of Mn^{II} in solid polycrystalline diamagnetic hosts in which the structure is better resolved than in frozen aqueous solution (Shaffer et al. 1976): the g-value is isotropic and very close to g_e. Since D is small, six transitions are possible between subsequent m_S levels; however, the spectrum is dominated by the $m_S = +1/2 \leftrightarrow -1/2$ transition. The latter consists of six hyperfine lines each split by a small anisotropy induced by the axial zero-field splitting D; in between these six lines there are five pairs of weak lines from forbidden $\Delta m_I = \pm 1$ transitions with an order of magnitude lower intensity than the main lines; this whole $m_S = \pm 1/2$ spectrum is on top of a very broad, rather structureless feature that is the sum of all the other five $\Delta m_S = 1$ transitions (e.g., $m_S = -3/2 \leftarrow -5/2$). Analytical expressions for the resonance conditions of all these transitions have been deduced some time ago (ibidem), but they are yet to be applied in bioEPR, perhaps because analysis is hampered by lower resolution due to increased linewidth. At higher microwave frequencies (≈ 35 GHz and beyond) this complex situation dissolves and a simple, isotropic, six hyperfine line pattern remains.

5.7 INTEGER SPINS

In the previous chapter we have identified integer-spin systems ($S = n$) as the most challenging ones from an EPR viewpoint because in the biological systems common weak-field limit ($S*S >> S*B$) in X-band they form non-Kramer's doublets that can be split even in the absence of an external magnetic field. For half-integer spin systems we resorted to rhombogram analysis as an alternative to the perturbation-theory development of analytical expressions for resonance conditions, but for integer systems, even this alternative proves to be of little value. The remaining options are to either limit oneself to a qualitative "fingerprint" description, or to go for a full-blown QM matrix diagonalization analysis. Here, we give an outline of the qualitative approach.

Figure 5.13 is the integer-spin equivalent of half-integer spin in Figure 5.9; for $S = 1$ through 4 it gives the effective g-values of intradoublet transitions for axial symmetry ($E = 0$). Equivalent to Equations 5.27–5.30 for half-integer spins for this relatively simple case (which senso stricto never occurs in biology), analytical expressions can still be written out:

$$g_{\parallel}^{eff}(\pm m_S) = 2|m_S|g_{\parallel} \tag{5.39}$$

$$g_{\perp}^{eff} = 0 \tag{5.40}$$

However, the effect of introducing a finite rhombicity is very different from that for half-integer spins in Figure 5.12: the non-Kramer's doublets are actually split by the rhombic zero-field interaction, and this results in a change in $g_{\parallel}^{eff}$ in the direction of *higher* values, that is, in the direction of lower fields. For pronounced rhombicities all $g_{\parallel}^{eff}$ "disappear" in infinity (zero field); in other words, the rhombicity-induced

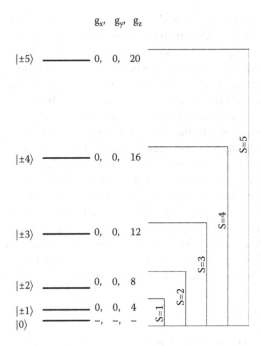

FIGURE 5.13 Effective g-values for integer spin systems in axial symmetry. The scheme gives the g-values for all transitions within the non-Kramer's doublets of $S = n$ systems assuming $g_{real} = 2.00$ and $S*S \gg S*B$; cubic zero-field terms are ignored.

splitting in the non-Kramer's doublet becomes greater than the X-band microwave energy, and transitions become impossible.

An additional complication for integer spin system is the significant occurrence of higher-order spin–spin interactions. Thus far we have represented the zero-field interaction in shorthand notation as $S*S$, which suggests that it is an interaction quadratic in the spin S. EPR theory actually allows the occurrence of higher-order terms (Abragam and Bleaney 1970: 140) that we could write in shorthand as $S*S*S$, $S*S*S*S$, etc., or perhaps as $S*^3$, $S*^4$, etc., and the only limitation is the magnitude of the system spin S: an $S*^n$ interaction is allowed (i.e., likely to be of finite strength) when $n \leq S/2$. Furthermore, in contrast to the E-term, which requires the symmetry of the paramagnet to be less than axial, $S*^n$ terms are not restricted by symmetry (in physics language, they describe a cubic interaction), and so they will always be present. Also, these terms induce a splitting of the non-Kramer's doublets, which not only means that they are never fully degenerate, but also that we have to consider the combined effect of the rhombic E-term and the higher-order cubic terms. The overall result turns out to be remarkably simple.

The rhombic zero-field splitting in non-Kramer's doublets is more pronounced the lower the $\pm m_S$ value of the doublet. For example, for $S = 2$ the zero-field splitting of the ± 1 doublet is $6E$, while that of the ± 2 doublet is $3E^2/D$ (ibidem: 212), and obviously $3E^2/D \ll 6E$ because $E/D < 1/3$. Consequently, the effective $g_\parallel$ of lower

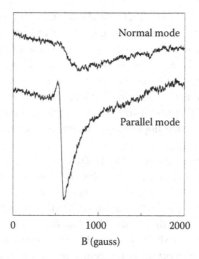

FIGURE 5.14 Dual-mode $S = 2$ EPR from an iron protein. The $T = 9$ K spectra are from a mononuclear high-spin ferrous site in dithionite-reduced desulfoferrodoxin from *Desulfovibrio vulgaris*. The top trace was recorded in normal or perpendicular mode ($B_1 \perp B_0$); the bottom trace was taken in parallel mode ($B_1 \parallel B_0$). (Modified from Verhagen et al. 1993.)

doublets will more rapidly diverge towards infinity for increasing E, and even a small *distribution* in the E-value due, for example, to protein conformational distribution, will lead to excessive line broadening.

An example is the $S = 2$ spectrum from a mononuclear high-spin Fe^{II} site in the protein desulfoferrodoxin given in Figure 5.14. A relatively sharp effective feature is observed from within the $m_S = \pm 2$ non-Kramer's doublet, but the g_z-value is shifted from the axial $g_{\parallel} = 8$ value (cf. Equation 5.39) to a higher g-value (i.e., lower field). Also, an extremely broad feature from the $m_S = \pm 1$ doublet, expected at $g_z \geq g_{\parallel} = 4$, extends over the whole field range of the figure.

5.8 INTERPRETATION OF G, A, D

The EPR spectrum is a reflection of the electronic structure of the paramagnet. The latter may be complicated (especially in low-symmetry biological systems), and the precise relation between the two may be very difficult to establish. As an intermediate level of interpretation, the concept of the spin Hamiltonian was developed, which will be dealt with later in Part 2 on theory. For the time being it suffices to know that in this approach the EPR spectrum is described by means of a small number of parameters, the spin-Hamiltonian parameters, such as g-values, A-values, and D-values. This approach has the advantage that spectral data can be easily tabulated, while a demanding interpretation of the parameters in terms of the electronic structure can be deferred to a later date, for example, by the time we have developed a sufficiently adequate theory to describe electronic structure. In the meantime we can use the spin-Hamiltonian parameters for less demanding, but not necessarily less relevant applications, for example, spin counting. We can also try to establish phenomenological (i.e., lacking a theoretical explanation) relationships between the parameters and some known structural feature, for example, between g-values and axial ligands to heme groups. Or we can take a semiquantitative approach by taking reasonably accurate analyses of electronic structures of well-defined model compounds, usually of high symmetry, and apply these to less well defined, complex structures. To this goal it is useful to develop a general feel for what the different spin Hamiltonian parameters represent in terms of chemical bonding. For transition ion complexes, for example, metalloproteins, we use simple crystal-field pictures with hydrogen-like wavefunctions.

A free electron in a magnetic field has the free-electron g-value $g_e = 2.0023$. When the electron is placed in an atomic or a molecular structure, it will orbit around one or more atoms, and so it will have orbital angular momentum. Its g-value deviates from g_e, and this deviation reflects the orbiting. All biologically relevant transition metals are open-shell d-ions, that is, they have an incomplete (less than ten) number of d-electrons in their outer shell. The d-orbits have directionality (they point to a direction in space from the origin at the center of the metal atom). Orbiting can be envisioned as the unpaired electron being in a specific d-orbital (its ground state), but every once in a while jumping up in energy to another d-orbital with a different directionality, thus sampling 3-D space around the atom. The further this excited state is separated from the ground state, the more difficult is the orbiting, and thus the smaller the deviation from the free electron g-value: g-anisotropy is a measure of energy separations between electronic ground and excited states.

The d-orbitals are labeled d(z^2), d(x^2-y^2), d(xy), d(xz), and d(yz), and in an ML_6 complex they subdivide into two groups depending on whether their angular dependent part points *along* molecular axes (the first two) or *in between* molecular axes (the last three). These groups are commonly labeled the E_g and T_{2g} set, following a notation from groups theory, and the individual orbitals are labeled e_g (2×) and t_{2g} (3×). In a simple crystal-field picture, the ligands of a coordination complex are represented as negative point charges. In a regular octahedral complex we have a metal ion surrounded by six identical ligands on the molecular axes at equal distances from the metal as shown in Figure 5.15a. Electrostatic interaction between the d-electrons and the ligands is repulsive, and is strongest for those d-orbitals that are along the molecular axes, that is, those that point directly towards the ligands. This results in the electron energy scheme of Figure 5.15b. When the two ligands on the z-axis are slightly removed away from the complex (i.e., elongation of the octahedron; Figure 5.15c), then all d-orbitals with z-character, d(z^2), d(xz), and d(yz), become lower in energy respective to the remaining two orbitals, d(x^2-y^2), d(xy), as shown in Figure 5.15d. Any additional deformation not along the z-axis will also lift the degeneracy between d(xz) and d(yz). This simple picture suffices for our qualitative interpretation of g-values in bioEPR, which we illustrate with two examples.

As a model system we take the hydrated cupric ion $Cu^{II}(H_2O)_6$, which is an elongated CuO_6 octahedron, so the electron energy scheme of Figure 5.15d applies, and the EPR spectrum is given in Figure 3.4. The electronic ground state of the Cu^{II} ion is $3d^9$ (or [Ar]$3d^9$, or $1s^2 2s^2 2p^6 3s^2 3p^6 3d^9$); that is, there are nine d-electrons to be placed in the scheme of Figure 5.15d, and with the Pauli exclusion principle ("maximally two paired electrons in an orbital") the result is that of Figure 5.15e. The unpaired

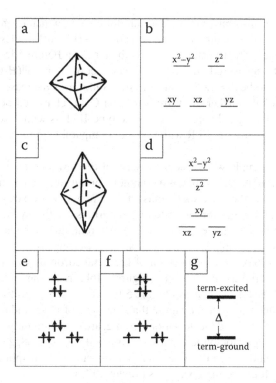

FIGURE 5.15 Crystal-field pictures for the d^9 system in deformed octahedra. (a) regular octahedron for a central metal surrounded by six identical ligands; (b) relative energy of d-orbitals for an octahedral complex; (c) octahedron elongated along the z-axis; (d) d-orbitals for an elongated octahedron; (e) ground state electron configuration for d^9; (f) excited state electron configuration for d^9 (g) term scheme for the configurations in (e) and (f).

electron is found in the $d(x^2-y^2)$ orbital. In order for it to have orbital angular momentum (to orbit around the Cu nucleus) it must sometimes be in a d orbit with different directionality. To achieve this, an electron must be promoted from either the $d(xz)$ or the $d(yz)$ to the $d(x^2-y^2)$ as in Figure 5.15f. Another way of expressing this is to say that spatial overlap between the $d(x^2-y^2)$ and, for example, the $d(xz)$ orbital allows the unpaired electron in the $d(x^2-y^2)$ orbital to "gain" some $d(xz)$ "character" (and therefore to lose some of its $d(x^2-y^2)$ character) by means of spin-orbit coupling. The electron scheme in Figure 5.15f is energetically less favorable than that in Figure 5.15e, and we can visualize this in a single picture by representing each scheme by a single, overall term energy to obtain the term scheme in Figure 5.15g. The smaller the difference between the terms, that is, the smaller the single-electron difference $\Delta[\, d(x^2-y^2)-d(xz)]$, the easier the orbiting, and thus the greater the deviation of g from g_e. This simple model affords the expressions (Bleaney et al. 1955)

$$g_{\parallel} \simeq g_e(1-8\lambda/\Delta) \qquad (5.41)$$

$$g_{\perp} \simeq g_e(1-2\lambda/\Delta) \qquad (5.42)$$

in which λ is the spin-orbit coupling constant for copper, which is -830 cm^{-1} for the free Cu^{II} ion, but whose value reduces typically to -710 cm^{-1} in these type of coordination complexes. The transition between the terms in Figure 5.15g is a d–d transition, and can be measured with optical spectroscopy. For Cu^{II}(H$_2$O)$_6$ this energy splitting is found to be $\Delta \approx 12300$ cm^{-1} or circa 800 nm. A *decrease* in Δ will cause the g-values to move further away from g_e. The CuO$_6$ elongated octahedron is an initial model for biological copper sites of the type-II class with hard-ligand (O, N) coordination and near-axial EPR with g-values comparable to those of the hydrated copper ion.

As a second example we consider the case of the low-spin ferric ion with electronic ground state [Ar]3d^5, which is a common configuration in hemoproteins. The low spin implies that there are six ligands (cf. Figure 5.3) and we can use exactly the same crystal-field scheme for elongated octahedra as for the hydrated copper ion. The resulting EPR spectra are, however, very different, due to the large value of the main crystal-field splitting, Δ, resulting in all five d-electrons to be confined to the T$_{2g}$ set (Figure 5.16a). Axial elongation of the octahedron plus a finite rhombicity makes all the t$_{2g}$ orbitals inequivalent (Figure 5.16b), resulting in three configurations (5.16c) and the term scheme in Figure 5.16d. The energy splittings between the terms are now much smaller (i.e., more than an order of magnitude) than for the d^9 picture in Figure 5.15g, and the deviations from g$_e$ are concomitantly much more pronounced. For example, for cytochrome c (Figure 5.4F) $g_{zyx} = 3.08, 2.23, 1.24$ (Hagen 1981). The relation between the axial and rhombic crystal-field splittings within the T$_{2g}$ set (Figure 5.16b) and the g-values is (Taylor 1977)

$$\Delta_{rhom} / \lambda = \frac{g_x}{g_z + g_y} + \frac{g_y}{g_z - g_x} \tag{5.43}$$

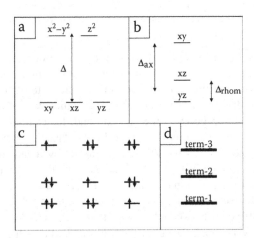

FIGURE 5.16 Crystal-field pictures for the low-spin d^5 system in deformed octahedra. (a) relative energy of d-orbitals for an octahedron with strong crystal field; (b) d-orbitals of the T$_{2g}$ set in a rhombically distorted octahedron; (c) three possible electron configurations of increasing total energy; (d) the term scheme for the configurations in (c).

$$\left(\Delta_{ax} + \Delta_{rhom}/2\right)/\lambda = \frac{g_x}{g_z + g_y} + \frac{g_z}{g_y - g_x} \tag{5.44}$$

in which λ is the spin-orbit coupling constant for Fe^{III}; for the free ion $\lambda \approx 420$ cm^{-1} (Harris 1966). We can determine the two crystal field splittings from the three g-values, but not the other way around. For cytochrome c we get: $\Delta_{ax} \approx 1100$ cm^{-1} and $\Delta_{rhom} \approx 600$ cm^{-1}. The optical d–d transitions corresponding to Δ_{ax} and Δ_{rhom} fall in the infrared, and have never been measured.

How to link these crystal-field parameters in a rational manner to the local molecular structure may not be obvious, but one can try one's luck in a phenomenological research, for example, by comparing Δ-values obtained from the EPR spectra of heme proteins with known structure, and hoping to discover some statistical connection between the two. W.R. Blumberg and J. Peisach have been the champions of this approach by constructing plots of Δ_{ax} versus Δ_{rhom} and relating areas to specific combinations of axial protein ligands, for example His/Met in cytochrome c (Brautigam et al. 1977, and references quoted herein). These plots are informally widely known as "truth tables," which I suspect is a pun by the authors to remind us of the lack of a theoretical basis and of the inherent corollary that results in the past are no guarantee for future predictive power, as illustrated later by the failure of this approach for the case of Met/His ligation (Teixeira et al. 1993).

A comparable attempt has been made for copper protein by plotting $g_{||}$-values versus $A_{||}$-values, which readily leads to the division into type I ("blue") and type II ("non-blue") copper proteins corresponding to small $A_{||}$-values (40–90 gauss) and an intense blue color versus greater $A_{||}$-values (120–220 gauss) and a very weak blue color (Malmström and Vängaard 1960, 1968), but which has subsequently been explored for statistically significant groupings in terms of the "main" coordinating atoms, that is, the equatorial ones in a structure presumed to be approximately an elongated octahedron (Peisach and Blumberg 1974). In the intervening years a wealth of structural information from copper protein crystallography has been obtained, and the frequently recurring theme of unexpected coordinations usually of rather low symmetry has turned the copper $g_{||}/A_{||}$ truth table into somewhat of a relic of historical interest.

Comparing relative values for hyperfine coupling constants does, however, give us a qualitative feel for "spin density," that is, for an answer to the question: "Where is the unpaired electron?" For one, detecting *ligand* hyperfine structure in a spectrum from a transition ion complex (as in Figure 5.6) is a direct observation of covalency, that is, of the fact that the unpaired d-electron of the metal spends part of its time on the ligand(s). The mirror image of this experiment is the observation of a reduced central hyperfine interaction with the metal nucleus: type-I blue copper proteins have small $A_{||}$-values because the unpaired electron has considerable spin density on the sulfur ligand (Fittipaldi et al. 2006), which unfortunately does not have a nuclear spin.

In a similar vein, the observation of multiple hyperfine interactions in the spectra of organic radicals allows for a delineation of the distribution in space and time of the unpaired electron over the atoms of the compound. A helpful simple relation, known as the McConnell equation, comes from the observation of isotropic hyperfine splitting from six equivalent protons in the symmetrical π-radical $C_6H_6^{\cdot-}$:

$$\rho_C = -(A_H)^{-1} \times 22.5 \text{ (gauss)} \tag{5.45}$$

in which ρ_C is the spin density on a particular C-atom in an aromatic ring system and A_H is the hyperfine splitting from the α-proton in the C-H fragment. The equation is based on the observation that in the $C_6H_6^{\cdot-}$ radical each of the six protons gives a splitting of 3.75 gauss, and on a theory in which aromatic C-H fragments hold an unpaired π-electron and two σ-CH bonding electrons (McConnell and Chesnut 1958). An analogous expression was subsequently proposed for β-protons of alkyl radicals (Heller and McConnell 1960):

$$\rho_C = -(A_H)^{-1} \times (C_0 + C_2 \cos^2 \theta) \tag{5.46}$$

in which typically $C_0 \approx 1$ gauss and $C_2 \approx 42$ gauss, and θ is the dihedral angle between the radical π-orbital and the C-H bond of the β-proton such that $\theta = 0$ corresponds to the eclipsed configuration. This relation has been widely used in bioEPR, for example, for amino-acid-derived radicals (Jeschke 2005) such as the tyrosyl radical in ribonucleotide reductase (Sjöberg et al. 1978) or the allylic radical in several dehydratases and reductases (e.g., Kim et al. 2008, and references therein).

A simple interpretation of the values of the zero-field interaction parameters (D, E, and higher-order coefficients) is probably the least straightforward. In early work on the Mn^{II} $S = 5/2$ system spin-orbit coupling was concluded to be dominant out of a range of considered mechanisms to give rise to zero-field interaction (Sharma et al. 1966, 1967, 1968). There is no direct parallel with the g-value deviation from g_e because here, also, couplings to excited states of different multiplicity ($S \neq 5/2$) have to be considered. The second most important mechanism was suggested to be spin–spin interaction, but its fractional contribution remained unclear (ibidem), and the matter appears to have remained controversial even for small model compounds up until this day (e.g., Neese 2003, 2006, 2007). Also, truth tables for high-spin systems based on zero-field splitting parameters have never appeared. What remains is a loose association of D-values with certain classes of complexes (e.g., near-octahedral Mn^{II} has $D \approx 0.1$ cm^{-1}; heme Fe^{III} has $D \approx 10$ cm^{-1}) and the necessity to determine the values of D and E for an understanding of the shape of EPR spectra and (cf. Chapter 12) for the temperature dependence of high-spin signals. Later, in Chapter 11 we will see that zero-field terms can also arise from the interaction between two or more paramagnets (dimers, clusters, organic triplets), and then their magnitude is in principle readily interpreted in terms of interaction strength, which is a function of distance and mutual orientation of interacting couples.

6 Analysis

Molecular spectra can be analyzed for spectro*metric* or for spectro*scopic* purposes. The term "spectrometric" usually refers to compound identification (linking a signal to a known structure) and to the determination of its concentration. The term "spectroscopic" stands for interpretation of the spectrum in terms of structure (chemical, electronic, nuclear, etc.). In this chapter we will look as some theoretical and practical aspects of a key spectrometric application of bioEPR, namely, the determination of the concentration of paramagnets, also known as spin counting. Subsequently, we consider the generation of anisotropic powder EPR patterns in the computer simulation of spectra, a basic technique that underlies both spectrometric and spectroscopic applications of bioEPR.

6.1 INTENSITY

A foremost strength of bioEPR spectroscopy is its applicability as an analytical chemical instrument for the determination of concentrations of prosthetic groups and of their stoichiometries in systems ranging in complexity from single proteins to whole cells. This particular strength clearly discriminates bioEPR from optical spectroscopy, which suffers not only from the practical problem of opacity of complex biological preparations, but also from the fundamental problem of an a priori undetermined wavelength-dependent intensity (or transition probability). In other words, the extinction coefficients of an optical absorption spectrum have to be determined experimentally, while all "extinction coefficients" of an EPR spectrum are equal to unity by definition: the intensity of an EPR spectrum follows directly from theory (i.e., from an interpretation of the spectrum in terms of its EPR parameters), and concentrations are straightforwardly determinable in terms of the molarity of an external EPR standard of known concentration. In mathematical terms, quantitative optical spectroscopy is based on Beer's law:

$$I(\lambda) = \varepsilon(\lambda)cd \tag{6.1}$$

in which the spectral amplitude I is a function of an unknown set of molar extinction coefficients ε, of the concentration of the chromophore c, and of an experimentally adjustable optical path length d, and in which other experimental conditions (light intensity, slit width, photomultiplier response function, etc.) are implicitly assumed to be standardized. Similarly, the amplitude of an EPR spectrum depends linearly on the concentration of the paramagnet, c, the sample tube diameter, d, and on a number of other experimental conditions (temperature, modulation amplitude, etc.) assumed to be standardized (see below). However, the EPR equivalent of $\varepsilon(\lambda)$ is a known quantity, which has been formulated in the literature in several equivalent forms (Holuj 1966; Pilbrow 1969; Isomoto et al. 1970; Abragam and Bleaney 1970: 136), for example,

$$I(l_i) = [g_x^2 g_y^2(1-l_z^2) + g_y^2 g_z^2(1-l_x^2) + g_z^2 g_x^2(1-l_y^2)] / 2g^2 \tag{6.2}$$

or

$$I(l_i) = \sum_{i=x,y,z} (g_i^2 - g^{-2}l_i^2 g_i^4) \tag{6.3}$$

in which the l_i's are the direction cosines that we defined in Equation 5.3 and g is the anisotropic g-value of Equation 5.5 or 5.6. The equivalent Equations 6.2–6.3 are yet other examples of expressions of limited practical value because they are valid for frequency-swept spectra only. For field-swept spectra of effective or real $S = 1/2$ systems, the intensity has to be divided by $g(l_i)$ (Aasa and Vänngård 1975), and so, for example, Equation 6.3 becomes

$$I(l_i) = g^{-1} \sum_{i=x,y,z} (g_i^2 - g^{-2}l_i^2 g_i^4) \tag{6.4}$$

in which g, or $g(l_i)$, is defined in Equation 5.5. This division by g is informally known as "the Aasa correction factor." To determine the spin concentration of a paramagnet from its powder spectrum Equation 6.4 should be integrated over space (cf. Section 6.3 to follow), but in practice an approximating average is used (ibidem):

$$I \approx \frac{2}{3} \sqrt{\frac{g_x^2 + g_y^2 + g_z^2}{3}} + \frac{g_x + g_y + g_z}{9} \tag{6.5}$$

When the (effective or real) g-values can be read from the spectrum (with Equation 2.6) then the factor I is known, and the EPR equivalent of Beer's law at fixed microwave frequency, v, has no unknowns except for the concentration c

$$I_v = Icd \tag{6.6}$$

which can then be determined by comparison with the EPR of *any* (effective) $S = 1/2$ compound of known concentration.

6.2 QUANTIFICATION

We already noted that EPR spectrometers usually measure the first derivative of the EPR absorption with respect to the magnetic field. Integration of the EPR spectrum (in practice, usually numerical integration of the digitized spectrum) affords the EPR absorption spectrum, and a subsequent integration (i.e., second integral or double integral: $\iint$) gives the area under the EPR absorption spectrum. The latter gives us a relative measure for the concentration of the paramagnet, provided a correction is made for the anisotropic transition probability over the powder pattern, I, that we have just defined in Equation 6.4 or closely approximated in Equation 6.5.

The procedure of "spin counting" is then to use the EPR spectrum of another paramagnetic compound as an *external* standard (which we will label "K" to avoid confusion with the spin S) of known concentration (c_K) to obtain the unknown (U) concentration (c_U) of the paramagnetic compound of interest as

$$c_U = c_K \left(\iint_U / \iint_K \right) (I_K / I_U)$$

(6.7)

For Equation 6.7 to be valid it is assumed that all other experimental conditions are equal for the two samples. If this is not true, additional corrections may be required for differences in modulation amplitude (M), microwave power attenuation in |dB| (P), magnetic field scan width (W) (or equivalently, the step width in gauss between two subsequent digitization points), electronic gain (G), sample diameter (d), and absolute temperature (T):

$$\frac{c_U}{c_K} = \frac{\iint_U}{\iint_K} \frac{I_K}{I_U} \frac{M_K}{M_U} 10^{\frac{P_U - P_K}{20}} \left(\frac{W_K}{W_U} \right)^2 \frac{G_K}{G_U} \frac{d_K}{d_U} \frac{T_U}{T_K}$$

(6.8)

The choice of the standard compound is in principle arbitrary; a common practical choice is the $S = 1/2$ system: 1-10 mM $CuSO_4/10$ mM $HCl/2M$ $NaClO_4$. The 10 mM HCl (i.e., *not* pH 2) is the right proton concentration to get a single-component Cu(II) spectrum (Figure 3.4). The "nonligand" perchlorate in high concentration is to prevent aggregation of copper. The resulting complex is thought to be the "Jahn–Teller distorted" (elongated along a molecular axis) axial $Cu^{II}(H_2O)_6$, that is, an elongated octahedron with O atoms at the six vertices and a Cu atom in the center; its g-values are $g_\parallel = 2.404$, $g_\perp = 2.076$; $A_\parallel$ (^{65}Cu) = 131 gauss and $A_\perp$ is unresolved (Hagen 1982a).

The spectrum of the CuO_6 elongated octahedron is in Figure 6.1 (we looked at it earlier on an extended field range in Figure 3.4). The figure also shows the first integral, that is, the EPR absorption spectrum, and the second integral, that is, the integrated EPR absorption spectrum. Note that, *ideally*, both the first-derivative EPR spectrum and the EPR absorption spectrum should start and end with zero intensity, but that the second integral starts with zero and ends with a finite constant value. This is the numerical value of $\iint_K$ in Equations 6.7 and 6.8. Before integration, we first "null" the spectrum, that is, we draw a straight (not necessarily horizontal) baseline from the first to the last point of the spectrum. A straight sloping baseline added to or subtracted from the derivative EPR spectrum simply means that the EPR absorption spectrum in its entirety is lifted up or down, that is, its offset is adjusted. The previously italicized "ideally" signals us to be on guard for common practical problems that can arise from baselines that are not straight, and spectra that can be deformed (e.g., by interference with overlapping signals).

An illustrative example is given in Figure 6.2: the spectrum (Arendsen et al. 1993) of a high-spin ferric heme with its typical very wide pattern of an intense feature around $g^{eff} \approx 6$ and a weak negative peak near $g \approx 2$ (Equations 5.34 and 5.35).

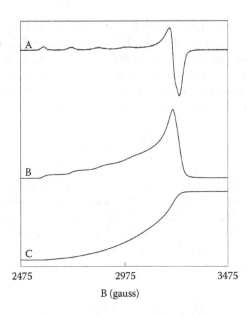

FIGURE 6.1 Integration of an EPR spectrum. The EPR derivative spectrum of the hydrated copper ion (trace A) is numerically integrated to its EPR absorption spectrum (trace B) and a second time integrated (trace C) to obtain the area under the absorption spectrum. Note that both the derivative and the absorption spectrum start and end at zero, while the doubly integrated spectrum levels off to a constant value: the second-integral value.

At first sight, the spectrum in trace A appears to be of rather acceptable quality with a slightly sloping baseline and a minor contaminant around $g \approx 2.4$ (circa 2750 gauss) from low-spin heme. However, a straightforward integration from left to right (trace B) amplifies the interferences, and the resulting double integral value is negative, which is physically a nonsensical result. The situation can be improved by judicious selection of integration limits (i.e., by starting the integration just before the first feature of the high-spin spectrum, and ending just after the last feature as indicated by the arrows in trace C). This eliminates the detrimental effects on the integral of long stretches of irregular baseline outside the powder pattern of interest, however, an irregular underlying baseline and also overlapping contaminating signals are not eliminated. The resulting first integral in trace C still exhibits some deformation (e.g., it dives below the zero base line in the middle of the spectrum), and in this example the resulting value for the double integral is only circa 50% of the actual high-spin heme concentration. In other words, the spin counting underestimates the concentration by a factor of two. Part of this effect could have been eliminated by correcting the spectrum with a baseline from a buffer sample without protein taken under identical conditions, but this would not have eliminated the interfering low-spin signal. Another frequently occurring error is illustrated in trace D where the spectrum is integrated only up to a certain intermediate field value as a consequence of misinterpretation of the spectrum (e.g., because the weak g_z-line is obscured by a radical or an iron–sulfur signal). Now the first integral appears to be of high quality,

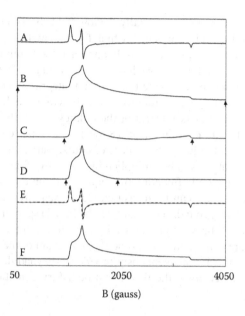

FIGURE 6.2 Interferences in spectral integration. Trace A is the experimental spectrum of a high-spin ferric heme, namely, siroheme in the dissimilatory sulfite reductase enzyme "desulforubidin" from *Desulfosarcina variabilis* (Arendsen et al. 1993). Trace B is the first integral of this spectrum illustrating the deleterious effect of an imperfect baseline on integration. Trace C also is the first integral but now computed over a limited field range indicated by the two arrows. In trace D the integration limits are further limited to cover only part of the heme spectrum. Trace E reproduces the experimental spectrum overlaid with a simulated spectrum, and trace F is the integral of the simulation.

but the resulting second integral value is again only circa 50% of that of the full nondeformed spectrum.

It is also possible to skip the integration altogether, namely, by numerical *simulation* of spectra and comparison of amplitudes. The quantitation is then intrinsic in the simulation which encompasses Equation 6.4 for the transition probability (see also below). This approach is indicated when experimental spectra are noisy and/ or disturbed by contaminating signals such as from dirty or damaged cavity walls, or from radicals in dye-mediated redox titrations, or from other centers in multi-center metalloproteins. In the latter case multiple-component simulations are particularly useful to determine stoichiometries of centers within a single metalloprotein, or indeed of centers in different metalloproteins in complex systems, for example, respiratory chains (Albracht et al. 1980). In Figure 6.2, trace E, the high-spin heme spectrum, has been simulated, and the integrated simulation is in trace F. The latter affords a value for the second integral that corresponds to a concentration close to 100% of the actual high-spin heme concentration.

It is well possible to do a spin counting when only part of the EPR powder spectrum is available, for example, because some features are broadened beyond detection or are at field values beyond the maximum limit of the magnet, or because the spectrum is disturbed by overlap of spectra from other paramagnets. Two conditions should be fulfilled: (1) one feature (usually the low-field absorption-shaped peak, cf. Figure 5.4E) should be "well separated," which for the low-field peak means that its tracing, on its high-field side, should closely approach the baseline and (2) all three g-values should be known. Expressions have been derived to relate the area under an isolated peak (i.e., first integral) to the total second integral of a reference spectrum (Aasa and Vänngård 1975), but in order to avoid errors of normalization it is perhaps easier to relate the single-peak integral of the unknown directly to the single-peak integral of a standard: in Equation 6.7 or 6.8 the $\int\int$'s are replaced by $\int$'s. In the case of the Cu(II) standard, the low-field peak is of course part of a four-line hyperfine pattern, and the intensity has to be multiplied by a factor of four.

The intensity expression in Equation 6.5 requires all three g-values to be known. Sometimes not all g-values can be measured experimentally, and they have to be estimated on theoretical grounds. For example, the Fe(III) spectra of low-spin hemo-proteins frequently exhibit very pronounced g-anisotropy to the extent that two of the three g-values are either at fields beyond the maximum of the magnet and/or are associated with features inhomogeneously broadened beyond detection. With only the highest g-value determined the theoretical boundary condition for low-spin d^5 systems with $3 < g_z < 4$

$$g_x^2 + g_y^2 + g_z^2 \simeq (2 + g_e)^2 = 16.02 \qquad (6.9)$$

(Griffith 1971) can be used to estimate I, e.g., by taking $g_x \approx g_y$ (de Vries and Albracht 1983).

For complex systems, notably in the multi-component EPR of respiratory chain complexes, it frequently happens that due to extensive overlap of spectra not even a single spectral feature is sufficiently isolated for single-integration purposes. In these situations one can base the quantification on the single integral of the first *half* of a single peak, however, this method usually brings along a significant uncertainty due to the fact that single peaks are commonly asymmetric and so integration up to the maximum amplitude value does not afford exactly half the value of integration of the full peak. A relatively modest example is shown in Figure 6.3, where integration based on the first half of the low-field peak of a [2Fe-2S]$^{1+}$ spectrum times two gives circa 120% of the actual spin count.

6.3 WALKING THE UNIT SPHERE

Biochemical EPR samples are almost always collections of randomly oriented mol-ecules: (frozen) aqueous solutions in which each paramagnetic molecule points in a different direction. In order to generate simulations of these "powder" EPR spectra we have to calculate the individual spectrum for many different orientations and then add these all up to obtain the powder pattern. Numerical procedures that generate sufficient spectra to approximate a powder pattern are collectively known as "walk-ing the unit sphere" algorithms. Here is the basic procedure:

Figure 6.4 shows the magnetic field vector B in the molecular Cartesian axes system xyz whose orientation is defined by the polar angles θ (between B and z) and ϕ (between the projection of B on the x-y plane and x). The third dimension of the

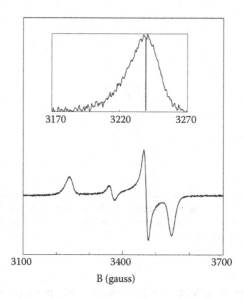

FIGURE 6.3 Quantification on the first half of an isolated peak. The spectrum is from the [2Fe-2S] cluster in the enzyme adenosine phosphosulfate reductase from *Desulfovibrio vulgaris* (Verhagen et al. 1993). The inset shows the asymmetrical low-field g_z-feature; the vertical line at the peak position indicates the rightmost integration limit for quantification on half a peak.

polar coordinates is, of course, the radius r along B. We want to sample a representative number of molecular orientations in the magnetic field, which is equivalent to the sampling of a representative number of orientations of the B vector in the xyz molecular axes system. To this goal we define a sphere whose center is at the origin of the xyz system and whose radius is equal to the length B. The vector B then becomes a unit vector touching the surface of the sphere. Our task is now to divide the surface of the sphere into a large number of equal subsurfaces and to let B sample a large number of orientations by pointing it to the middle of each of these little surfaces. This is mathematically equivalent to dividing up the sphere in a large number of cones of equal solid angle, and since the surface of a cone of angle 2θ is obtained from the integration

$$\int_0^{2\pi}\int_0^{\pi} \sin\theta\, d\theta\, d\varphi = \int_0^{2\pi}\int_{-1}^{1} d\cos\theta\, d\varphi \qquad (6.10)$$

it follows that we have to make a large number of equidistant steps in dcosθ for $0 \le \theta \le \pi$ (i.e., $1 \ge \cos\theta \ge -1$) and a (not necessarily identical) large number of equidistant steps in ϕ for $0 \le \phi \le 2\pi$. The overall field-swept EPR powder absorption spectrum $S(B)$ is obtained by discrete numerical integration:

$$S(B) = \int_0^{2\pi}\int_{-1}^{1} I[g(\theta,\varphi)]F\{B_{res}[g(\theta,\varphi)]\}d\cos\theta\, d\varphi \qquad (6.11)$$

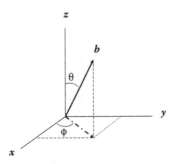

FIGURE 6.4 Vectors to describe a walk on the unit sphere. The orientation of a vector b of unit length along the dipolar magnetic field vector B in a Cartesian molecular axes system xyz is defined by the two polar angles: θ between b and the z-axis, and φ between the projection of b on the x-y plane and the x-axis.

in which I is the intensity defined in Equation 6.4, B_{res} is the resonance field (or line position) defined in Equation 5.4, and F is a lineshape function (e.g., the gaussian defined in Equation 4.8.)

A computer program to generate $S(B)$ in pseudo-code would be something along the following lines

```
INPUT: g-values g_i, linewidth Γ, frequency ν, field limits B_min-B_max
INPUT: stepwidth in solid angle dcosθ, dφ; spectral points n
NULL absorption-array
Compute stepwidth in field: dB = (B_max-B_min)/(n-1)
DO STEP in cosθ from 1 to -1 (θ from 0 to π)
    DO STEP in φ from 0 to 2π
        Compute direction cosines l_i(θ,φ) with Equation 5.3
        Compute g-value g(g_i,l_i) with Equation 5.5
        Compute intensity I(g) with Equation 6.4
        Compute resonance field B_res(g) with Equation 5.4
        DO STEP in dB from B_min to B_max
            Compute line shape F(B_res, Γ) with Equation 4.8
            ADD intensity F to absorption-array
        END STEP in dB
    END STEP in φ
END STEP in cosθ
```

This generic procedure affords the powder EPR absorption spectrum, which should be differentiated to get the powder EPR spectrum. Note that the whole procedure consists of three nested loops with the computation of an exponential (Equation 4.8) within the inner loop. Coded in a higher language (C, FORTRAN95) and run on a standard PC, this program will generate the EPR spectrum of a simple $S = 1/2$ or an effective $S = 1/2$ system in a split second (of the order of 10 ms or less). It is, however, useful to think about ways to make it as fast as possible, because extending the code for not so simple systems and adding options for (semi)automatic fitting of experimental data can increase overall computation time by many orders of magnitude. There are two key steps to significantly optimize the efficiency (or CPU time) of the procedure: (1) minimize the number of steps in the walk over the unit sphere, and (2) minimize the number of machine operations in the innermost loop.

The expression for the angular-dependent g-value Equation 5.5 is periodic over one octant of the unit sphere. Therefore, the walking steps can be limited to $1 \geq \cos\theta \geq 0$ and $0 \geq \phi \geq \pi/2$. The expressions for the intensity and the resonance field are functions of g, and so they have the same periodicity, and the overall computation time can be reduced by a factor of 8. Further time-reduction by a similar factor can be achieved by replacing the straightforward stepping in solid angle by more sophisticated schemes of probing the unit sphere. These schemes make use of the fact that for orientations close to $\theta = 0$ the angular-dependent expressions (e.g., Equation 4.8) are rather insensitive for changes in ϕ, and so the closer we are to $\theta = 0$, the less number of steps in ϕ suffice to approximate a powder pattern. One implementation of this idea is Belford's igloo scheme (Pilbrow 1990: 226) whose name derives from the fact that an igloo has only one ice block at its top but many at its base. In this scheme the number of steps made in ϕ increases linearly from one for $\cos\theta = 1$ (B along z) to a maximum for $\cos\theta = 0$ (B in the x-y plane).

Finally, the minimally required number of molecular orientations (steps in $\cos\theta$ and in ϕ) is determined "experimentally" by inspection of trial simulations as illustrated in Figure 6.5 on the now familiar high-spin heme spectrum: too few orientations cause so-called "mosaic artifacts," which must be eliminated by increasing the step numbers. In this particular example of the high-spin heme from Figure 6.2, the g_x and g_y-values are relatively close, and 25 steps in ϕ suffice, but g_z is well separated and the number of steps in $\cos\theta$ must be increased beyond 1000 to fully eliminate mosaic artifacts.

Note that mosaic artifacts can also occur *physically* in real spectra when a real powder sample of a model compound exhibits microcrystallinity and thus contains too few different molecular orientations. This phenomenon is rare in X-band EPR and is usually easily solved by grinding the sample in a mortar; it is, however, not at all uncommon even for extensively ground samples in high-frequency EPR with single-mode resonators where the sample size is orders of magnitude less than that of an X-band sample.

6.4 DIFFERENCE SPECTRA

We have seen in Chapter 2 that the frequency of an EPR spectrum is not a choice for the operator (once the spectrometer has been built or bought) as it is determined by the combined fixed dimensions of the resonator, the dewar cooling system, and the sample. Even if standardized sample tubes are used and all the samples have the same dielectric constant (e.g., frozen dilute aqueous solutions of metalloproteins), the frequency will still slightly vary over time over a series of consecutive measurements, due to thermal instabilities of the setup. By consequence, two spectra generally do not have the same frequency value, which means that we have to renormalize before we can compare them. This also applies to difference spectra and to spectra

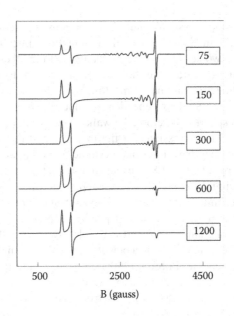

FIGURE 6.5 Mosaic artifacts in simulated spectra. The high-spin ferric heme spectrum from Figure 6.2 is simulated with 25 steps in polar angle φ and with an increasing number (75 to 1200) of steps in $\cos\theta$. The simulations are based on a "regular" grid (i.e., an equal number of φ-steps for each $\cos\theta$-value).

corrected by subtraction of a baseline spectrum obtained from an empty spectrometer or, better, from a duplicate sample that lacks the paramagnet but is otherwise (tube size, solvent, buffer, etc.) identical to the original one.

Suppose we want to compare two spectra—let's call them spectrum-α and spectrum-β—taken over field sweeps that may be identical but with a slight difference in their microwave frequency. The spectra are digital arrays corresponding to amplitudes at equidistant field values. The procedure to convert spectrum "β" taken at frequency v_β to frequency v_α of reference spectrum "α" is as follows: For each field value B_α of spectrum-α we calculate the corresponding field for v_β: $B_\beta = (v_\beta/v_\alpha)B_\alpha$, and then we search in spectrum-β to the two digital field values that nearly match (that "embrace") the value B_β in order to interpolate the two corresponding amplitudes to an intermediate amplitude value for B_β to be stored in a new array of β-amplitudes onto a B_α grid. In pseudo-code

```
INPUT: amplitude arrays α and β
INPUT: frequencies vα and vβ
INPUT: field limits Bα(start), Bα(end), Bβ(start), and Bβ(end)
CREATE array β-new
COMPUTE stepwidth dBα and dBβ
DO STEP in dBα from Bα(start) to Bα(end)
     Compute corresponding Bβ = (vβ/vα) Bα
     SEARCH β-array for two B-values that 'embrace' Bβ
     INTERPOLATE the corresponding two amplitudes using dBβ
     PUT the result in array β-new
END STEP in dBα
```

Formally, this procedure is correct only for spectra that are linear in the frequency, that is, spectra whose line positions are caused by the Zeeman interaction only, and whose linewidths are caused by a distribution in the Zeeman interaction (in g-values) only. Such spectra do exist: low-spin heme spectra (e.g., cytochrome c; cf. Figure 5.4F) fall in this category. But there are many more spectra that also carry contributions from field-independent interactions such as hyperfine splittings. Our frequency-renormalization procedure will still be applicable, as long as two spectra do not differ "too much" in frequency. In practice, this means that they should at least be taken at frequencies in the same band. For a counter-example, in Figure 5.6 we plotted the X-band and Q-band spectra of cobalamin (dominated by hyperfine interactions) normalized to a single frequency. To construct difference spectra from these two arrays obviously will generate nonsensical results.

Part 2

Theory

7 Energy Matrices

7.1 PREAMBLE TO PART 2

In this and the next two chapters we will develop some of the quantum mechanical background of EPR spectroscopy collectively known as the spin-Hamiltonian theory, and we will particularly be interested in specific adjustments, be they extensions or simplifications, that make the theory applicable to biomolecules. Although the theory is only a small subset of quantum mechanics in general, its understanding and subsequent application does require some proficiency in mathematics, notably regarding complex numbers and matrix algebra, and in the physics' view of wave equations and operators. Here, in contrast to our approach in previous chapters, very little will come "out of the blue"; effort has been put into assuring that each step made follows understandably from what preceded it and is amply illustrated by fully worked out examples. Having mastered the three chapters that make up Part 2 you will have acquired a profound understanding of EPR powder patterns of randomly oriented paramagnets, and you will be prepared to tackle just about any problem in continuous-wave biomolecular EPR spectroscopy. If, however, you decide now or during your reading of these three theoretical chapters that Part 2 is "a bridge too far," then simply skip it in the knowledge that many significant bioEPR experiments with chemical, biological, and/or medical bearing can be carried out with the background thus far provided in the previous chapters. Do, however, read on a little further into the next few paragraphs, which are intended to be a qualitative substitute and summary of the hard-core stuff later in Chapters 7–9.

According to the theory of quantum mechanics small particles, such as molecules (including, e.g., paramagnetic metalloproteins) can be described by means of a wavefunction (a mathematical expression containing goniometric functions like sines and cosines). "Described" here means that the intrinsic properties of the molecule (its geometric and electronic structure) are somehow stored in the coefficients of the sines and cosines of the wavefunction, analogous to the properties of a cell being stored in the base triplets of DNA. To us experimentalists, molecules only come to life when we let them "do something," that is, when they are subjected to certain external stimuli like a change in reactant concentration or a (change in) radiation, analogous to the (regulation of) DNA expression in a cell. Physicists call such an external stimulus an "operator" (an alternative name is a "Hamiltonian"), and they have developed an elaborate mathematical framework in which experimentally observing molecules is described as operators (Hamiltonians) working on (i.e., changing) wavefunctions. For spectroscopy, this approach has turned out to be extremely fruitful because the operator + wavefunction mathematics is an efficient way to determine all the energy levels of a molecule and, therefore, all the energy differences and thus all the wavelengths at which the molecule will absorb radiation, including the likelihood it will do this. In other words, this operator + wavefunction theory predicts all the lines and amplitudes of all the spectra of a molecule subject to a particular environment.

The bad news is that mathematically solving an operator + wavefunction problem for an average molecule in an average environment is generally so extremely complicated that one has to make very drastically simplifying assumptions to be able to obtain a solution (e.g., a prediction of a spectrum) within the life span of a human being. For EPR spectroscopy this means that we look at molecules with our eyes almost closed so that we only see the paramagnetic properties of the unpaired electron(s) of a molecule in its electronic ground state, and we describe these in very simple so-called "spin wavefunctions." The behavior of a particular unpaired electron in its specific environment, determined by other unpaired electrons (if present at all) and by a laboratory electromagnet, we describe by that very simple operator called a *spin Hamiltonian*. So the operator + wavefunction theory is simplified to a spin-Hamiltonian operator plus spin wavefunction theory. Of course, we cannot expect this "bare bones" theory to afford a full description of all the properties of the molecule. For example, we will not be able to predict all the spectra (from x-ray absorption to radio-frequency resonance) of the molecule, but we should at least be able to describe properties associated with the ground-state molecular paramagnetism (e.g., the EPR spectrum). This description will be in terms of EPR parameters like the ones we met in previous chapters: g, A, D, etc. Somehow these "spin parameters" are related to the details of the geometric and electronic structure of the molecule, but if we want to make any sensible statement about these details on the basis of EPR experiments, then we must go through the tedious procedure of "desimplifying" the operator + wavefunction theory to a more complex level of description, so that we can relate spin parameters like g, A, and D to, for example, electrons in molecular orbitals, including those of excited electronic states. In fact, in biomolecular EPR spectroscopy, we may not be particularly inclined to make this effort because we happen to be interested in more mundane pieces of information such as how many copper(II) ions there are in a particular enzyme.

Even for this mundane info it is necessary that we understand (can describe) the EPR spectrum in terms of the spin parameters of the molecule under study. For specific cases we have given resonance-condition expressions in the previous chapters. The crux of the matter treated in Part 2 is that there is a general approach to describing EPR spectra: it is always possible to write down a spin Hamiltonian and its associated spin wavefunctions for any paramagnet and to use these to deduce expressions for the EPR resonance condition (peak positions) and transition probability (amplitudes): that is, to predict EPR spectra. Frequently, the operator + wavefunction approach may not be solvable analytically but only numerically with the aid of a computer, but it always involves predicting an EPR spectrum on the basis of particular values for spin parameters. And if the prediction corresponds with the experiment (the simulation fits the spectrum), then we understand the spectrum in terms of a spin model: what the spin of the system is, what its g, D, etc., parameters are, and whether nuclear and/or other electron spins may be involved. Computer programs that implement this general approach can be run as a black box; it is not strictly necessary to know what the underlying theory is. The program generates an EPR spectrum based on a spin model, and if this simulation (reasonably) fits the experiment then the spin model is (reasonably) OK.

Subtleties enter the consideration of what is reasonable due to the intrinsic lack of symmetry of complex biomolecules. Physicists employ symmetry as a means to conceptually and mathematically simplify problems, and the quantum mechanics of the operator + wavefunction method heavily relies on symmetry. Unfortunately, this means that describing a low-symmetry paramagnetic biomolecule with the simple spin Hamiltonian, and simple spin functions may actually turn out to be not so simple after all. Further simplifications usually have to be made to keep the spin Hamiltonian approach practical for biomolecules, and to judge the validity of these simplifications requires insight in their theoretical background; this is an important part of what is discussed in this and the subsequent chapters. If you skip Part 2, it is advisable to be modest about interpretational skills regarding fine details of spectra; if you happen to notice 23 different minor inclinations and shoulders in an EPR powder spectrum, then try to stay on the safe side and refrain from interpreting them as reflecting the presence of 23 different types of biomolecules. It is rather more likely that there is just one type of molecule and that an understanding of the fine details of the spectrum requires increased theoretical efforts.

A final point concerns the concept of "conformation," a word that has invaded just about every corner of biochemistry. Two molecular conformations are two different sets of stable spatial coordinates of the atoms of a single type of molecule (that is, no change in the chemical formula). Think, for example of the water molecule whose ground state is in the shape of a "V" of H,O,H atoms with an angle of 109°. The H_2O molecule could also have its atoms in a straight line: H-O-H, which is a different conformation, albeit a very unlikely one. Drastic biomacromolecular conformational changes are common in nature and are at the basis of many biological processes, for example, the hormone-induced conformational change of a G-protein in signal transduction, or the proton-gradient enforced conformational change of the active site of the ATP syntase enzyme in oxidative phosphorylation. Here, we are interested in the more subtle phenomenon of conformational *distributions*; the relatively flexible structure of a biomacromolecule ensures that its ground state is not a single unique conformation, but rather a large collection of slightly different conformations corresponding to a very shallow distribution well in energy space. Since two conformations of a molecule correspond to two different structures in 3-D space, they should also correspond to two different electronic structures, and this trickles down into a difference in paramagnetism, and thus into the magnitude of the parameters describing the EPR spectrum. In other words, two different conformations of a paramagnetic molecule result, for example, in two different g-values. And in a distribution of many conformations, each one has its own g-value. Since the differences are subtle, the result is not a large collection of different EPR peaks, but a large collection of unresolved peaks, that is, a broadening of resonance lines. A formal description of this broadening is developed in Chapter 9 under the name of "g-strain," and low symmetry, or the lack of symmetry, once more plays a role of significance. The bottom line is that g-strain broadening can make (bio)molecular EPR powder lines quite asymmetric, and the experimentalist should be prepared to recognize these patterns for what they are: a reflection of the flexible nature of the molecule and not an indication of site multiplicity, i.e., not an indication of the presence of different molecules or different prosthetic groups.

7.2 MOLECULAR HAMILTONIAN AND SPIN HAMILTONIAN

Just like any spectroscopic event EPR is a quantum-mechanical phenomenon, therefore its description requires formalisms from quantum mechanics. The energy levels of a static molecular system (e.g., a metalloprotein in a static magnetic field) are described by the time-independent Schrödinger wave equation,

$$H\psi = E\psi \tag{7.1}$$

in which ψ is a function known as the wave function that contains information on all the intrinsic properties of the sample molecule (i.e., the electronic and nuclear structure), and H is an operator that contains all the interactions of the world to which the molecule under study is subject. Letting H work on ψ defines a so-called eigenvalue problem whose solutions are the eigenvalues E, the stable energy levels of the molecule. In other words the procedure of modifying ψ by the operator H is equivalent to multiplying ψ by a set of constants E that happen to be the possible energies of the molecule "at rest," that is, when it is not subject to interaction with time-varying electromagnetic radiation and/or subject to chemical change. This mathematical problem can be exactly solved for a few very simple systems only, and the most familiar one of these is probably the hydrogen atom in vacuo.

Since a metalloprotein in an EPR tube is significantly more complex than a hydrogen atom in vacuo, the wave equation has to be simplified to be practical, and a common approach in chemistry is that of the Born–Oppenheimer approximation, in which electronic and nuclear motions are decoupled, and the nuclear configuration is declared frozen. This view still leaves a coordination compound as an electronic system of intimidating complexity, which the solid-state physicists, who developed EPR in the 1950s (Abragam and Pryce 1951, Stevens 1952), have sought to tackle by introducing the concept of the spin Hamiltonian. This describes a system with an extremely simplified form of the Schrödinger wave equation that is a valid description *only* of the lowest electronic state of the molecule *plus* magnetic interactions. In this description the simplified operator, H_S, is the spin Hamiltonian, the simplified wave functions, ψ_S, are the spin functions, and the eigenvalues E are the energy values of the ground state spin manifold. The complete electronic structure of the molecule is then contained in a small number of parameters in H_S, and it thus becomes possible to describe the EPR spectrum of the electronic ground state of a molecule without having to specify its detailed electronic structure. Here is an approach that is, above all, a practical strategy: spectral data can be tabulated in a small number of spin-Hamiltonian parameters and their tedious theoretical chemical analysis can be deferred to a later, more suitable point in time, for example, when theoretical knowledge has sufficiently advanced or when computer hardware has become sufficiently efficient, to make a reasonably accurate analysis realistic. Note that the analysis of some electronic-structure problems of biological paramagnets has only recently become "doable," and that many other problems are still essentially beyond our present reach. Moreover, the bioEPR spectroscopist may well decide the research perspective will be focused on very different goals than in-depth electronic structure analyses (e.g., on attempts to determine thermodynamic and kinetic parameters of redox prosthetic groups). Rather than being an intermediate step in a full-blown quantum chemical journey, the spin-Hamiltonian analysis then becomes the terminus of the spectroscopy to be translated into biochemically relevant information.

This choice of perspective (biochemical versus quantum chemical) and the associated choice of depth (spin Hamiltonian versus full Hamiltonian) is a strategic one of considerable consequence. Our decision to typically not go beyond spectral analysis at the level of the spin Hamiltonian, on the one hand, does not restrict us in any way in terms of the spectroscopy (any spectrum is useful and available for biochemical interpretation), while on the other hand it positions us at the periphery of hardcore quantum mechanics. A few relatively simple tricks suffice to solve just about any spin Hamiltonian problem, and perhaps the only real challenge is to develop a feel for how extensive we should make the description (how many terms in the spin Hamiltonian should be developed) while retaining practical applicability and meaningfulness. The two main tricks that we need are (1) the use of spin operators in particular raising and lowering operators, and (2) the diagonalization of the energy matrix. We will now develop these tools, below, and subsequently we will turn to developing an intuition for what particular spin Hamiltonian one should use for a specific bioEPR problem and a feeling for where to put the limits of its information content beyond which the specter of overinterpretation lures.

First of all, recall (hopefully) that, based on the definition

$$i \equiv \sqrt{-1} \tag{7.2}$$

a complex number x is written as

$$x = a + ib \tag{7.3}$$

and the "complex conjugate" of x is

$$x^* = a - ib \tag{7.4}$$

so that xx^* is always a real number:

$$xx^* = a^2 + b^2 \tag{7.5}$$

Now let us rewrite the wave Equation 7.1 in what is known as a *Dirac's bracket notation:*

$$E = \langle \psi | H | \psi \rangle \tag{7.6}$$

Note that here "bracket" does not mean just any round, square, or curly bracket but specifically the symbols "⟨" and "⟩" known as the "angle brackets" or "chevrons." Then ⟨ψ| is called a "bra" and |ψ⟩ is a "ket," which is much more than a word play because a bra wavefunction is the complex conjugate of the ket wavefunction (i.e., obtained from the ket by replacing all i's by $-i$'s), and Equation 7.6 implies that in order to obtain the energies of a static molecule we must first let the Hamiltonian work "to the right" on its ket wavefunction and then take the result to compute the product with the bra wavefunction "to the left." In the practice of molecular spectroscopy |ψ⟩ is commonly a collection, or "set," of subwavefunctions |ψ$_i$⟩ whose subscript index i runs through the number n that is equal to the number of allowed static states of the molecule under study. Equation 7.6 also implies the "Dirac function" equality

$$\langle \psi_i \| \psi_j \rangle = \delta_{i,j} \tag{7.7}$$

which is a shorthand notation for

$$\langle \psi_i \| \psi_i \rangle = 1$$
$$\langle \psi_i \| \psi_j \rangle = 0 \tag{7.8}$$

which, in its turn, is a mathematical shorthand for the fact that all |ψ$_i$⟩'s taken together form a "complete orthogonal set" or a full description of the static molecule in terms of independent subwavefunctions (independent because no |ψ$_j$⟩ can be derived from any of its |ψ$_k$⟩ congeners). Also, note that the characterization of the static molecule by the |ψ$_i$⟩'s is not unique, because any linear combination |φ$_i$⟩'s of |ψ$_i$⟩'s

$$\left| \varphi_i \right\rangle = d_{1i} \left| \psi_1 \right\rangle + d_{2i} \left| \psi_2 \right\rangle + \cdots + d_{ni} \left| \psi_n \right\rangle \tag{7.9}$$

will also do (i.e., there is an infinite number of possible characterizations). Of course, the |ψ$_i$⟩'s can also be expressed as linear combinations of, for example, |φ$_i$⟩'s

$$\left| \psi_i \right\rangle = c_{1i} \left| \varphi_1 \right\rangle + c_{2i} \left| \varphi_2 \right\rangle + \cdots + c_{ni} \left| \varphi_n \right\rangle \tag{7.10}$$

In this admittedly extremely brief and hermetic summary of the quantum mechanics of static molecules, the key issue for us is that Equations 7.6 and 7.10 imply that, provided we know what the static molecular Hamiltonian H looks like and provided we can write down any set |φ$_i$⟩, we can always obtain all the molecular energies E (and therefore the molecular spectrum) by computing all n^2 terms

$$a_{ij} = \left\langle \phi_i \middle| H \middle| \varphi_j \right\rangle \tag{7.11}$$

then constructing the $n \times n$ "energy matrix" E from all these a_{ij}'s

$$E = \begin{pmatrix} a_{11} & \cdots & a_{1n} \\ \vdots & \ddots & \vdots \\ a_{n1} & \cdots & a_{nn} \end{pmatrix} \tag{7.12}$$

and finally diagonalizing matrix E to E_{dia} through an axes transformation by means of a rotation matrix R

$$E_{dia} = RER^{-1} \qquad (7.13)$$

because the elements on the diagonal of E_{dia} are identical to the energies E_i of the static molecule subject to the interactions described by the Hamiltonian H, from which we can calculate the spectral line positions as $h\nu = \Delta E$. Furthermore, R gives us the coefficients c_{ij} in Equation 7.10 from which we can calculate the transition probabilities or amplitudes for the spectral lines. Except for some very simple cases, Equation 7.13 cannot usually be solved analytically, but fortunately, very efficient open-source computer algorithms are available to obtain numerical solutions. What remains for the spectroscopists is to decide what H should look like for a particular molecular system subject to time-independent interactions, to write out a proper set of $|\varphi_i\rangle$'s, to put all this in a computer, and to hit the Enter key. In the next sections, let us then try to develop the bioEPR version of this procedure.

7.3 SIMPLE EXAMPLE: S = 1/2

In Chapter 5 we developed a picture of the three main players whose mutual interaction underlies essentially all EPR spectral features: man-made magnets B, electron spins S, and nuclear spins I, and we alluded to their pair-wise interactions with the freehand notation of a connecting asterisk, (e.g., $B*S$). In the quantum-mechanical picture of the spin Hamiltonian a man-made magnet is still a magnet, and its vectorial strength remains B, however, spins become spin *operators,* and they work on a particular subset of wave functions, namely spin functions. Let us then become specific and translate the extremely general Equation 7.1 (or 7.6) into a form that directly and practically applies to the EPR problem of a magnetic ground manifold described by a spin Hamiltonian, i.e., to a system that can occur in a relatively small number of discrete states whose relative energies have to do with spins and a magnetic field.

Here are the basic rules of the game: For a system with electron spin S, the known complete orthogonal set of $2S + 1$ wavefunctions is associated with the values m_S and is written as

$$\varphi_i = |m_S\rangle \qquad (7.14)$$

The spin Hamiltonian contains electron spin operators that are completely defined as follows

$$S_z|m_S\rangle = m_S|m_S\rangle$$
$$S_x = (1/2)(S_+ + S_-) \qquad (7.15)$$
$$S_y = (1/2i)(S_+ - S_-)$$

in which

$$S_+\left|m_S\right\rangle = \sqrt{S(S+1)-m_S(m_S+1)}\left|m_S+1\right\rangle$$
$$S_-\left|m_S\right\rangle = \sqrt{S(S+1)-m_S(m_S-1)}\left|m_S-1\right\rangle$$
(7.16)

In other words, if the spin Hamiltonian contains a spin operator S_z then letting H_S work on a spin wavefunction $\left|m_S\right\rangle$ produces the eigenvalue m_S, but if H_S contains an operator S_y or S_x then we first have to rewrite them in terms of the so-called ladder operators (or "raising and lowering operators," "step-up and step-down operators," "shift operators," or in certain contexts, "creation and annihilation operators") S_+ and S_- which, when working on the spin wavefunction $\left|m_S\right\rangle$ give an eigenvalue associated with a *different* wavefunction $\left|m_S+1\right\rangle$ or $\left|m_S-1\right\rangle$. The complete set of wavefunctions is orthogonal as defined in Equations 7.7 and 7.8 or in terms of our spin wavefunctions:

$$\left\langle m_S\|m_S\right\rangle = 1$$
$$\left\langle m_S\|m_{S+i}\right\rangle = 0$$
(7.17)

Our task is now to write out the spin Hamiltonian H_S, to calculate all the energy-matrix elements in Equation 7.11 using the spin wavefunctions of Equation 7.14 and the definitions in Equations 7.15–7.17, and to diagonalize the complete E matrix to get the energies E_i and the intensities of the transitions. We will now look at a few examples of increasing complexity to obtain energies and resonance conditions, and we defer a look at intensities to the next chapter.

Suppose we have an isolated system with a single unpaired electron and no hyperfine interaction. Mononuclear low-spin Fe^{III} and many iron–sulfur clusters fall in this category (cf. Table 4.2). The only relevant interaction is the electronic Zeeman term, so the spin Hamiltonian is

$$H_S = \beta B(g_x l_x S_x + g_y l_y S_y + g_z l_z S_z)$$
(7.18)

which we rewrite, using the shorthand notation $G_i \equiv \beta B g_i l_i/2$, as

$$H_S = 2G_x S_x + 2G_y S_y + 2G_z S_z$$
(7.19)

The starting orthogonal set of spin wavefunctions is

$$\varphi_1 = \left|+1/2\right\rangle$$
$$\varphi_2 = \left|-1/2\right\rangle$$
(7.20)

The effect of the individual spin operators (Equations 7.15 and 7.16) on these functions is

$$S_z \left| +1/2 \right\rangle = 1/2 \left| +1/2 \right\rangle$$

$$S_z \left| -1/2 \right\rangle = -1/2 \left| -1/2 \right\rangle$$

$$S_+ \left| -1/2 \right\rangle = \sqrt{\frac{1}{2}\left(\frac{1}{2}+1\right) - \left(-\frac{1}{2}\right)\left(-\frac{1}{2}+1\right)} \left| +1/2 \right\rangle = \left| +1/2 \right\rangle$$

$$S_- \left| +1/2 \right\rangle = \sqrt{\frac{1}{2}\left(\frac{1}{2}+1\right) - \frac{1}{2}\left(\frac{1}{2}-1\right)} \left| -1/2 \right\rangle = \left| -1/2 \right\rangle$$

$$S_x \left| +1/2 \right\rangle = \frac{1}{2}S_- \left| +1/2 \right\rangle = \frac{1}{2}\left| -1/2 \right\rangle$$

$$S_x \left| -1/2 \right\rangle = \frac{1}{2}S_+ \left| -1/2 \right\rangle = \frac{1}{2}\left| +1/2 \right\rangle$$

$$S_y \left| +1/2 \right\rangle = i\frac{1}{2}S_- \left| +1/2 \right\rangle = i\frac{1}{2}\left| -1/2 \right\rangle$$

$$S_y \left| -1/2 \right\rangle = -i\frac{1}{2}S_+ \left| -1/2 \right\rangle = -i\frac{1}{2}\left| +1/2 \right\rangle$$

(7.21)

and so the energy matrix for this set of wavefunctions is

$$E = \beta B \begin{pmatrix} \dfrac{1}{2}g_z l_z & \dfrac{1}{2}(g_x l_x - i g_y l_y) \\ \dfrac{1}{2}(g_x l_x + i g_y l_y) & -\dfrac{1}{2}g_z l_z \end{pmatrix}$$

(7.22)

or, in a slightly different notation, again using the shorthand $G_i \equiv \beta B g_i l_i / 2$,

$$\begin{array}{c} \langle +1/2 | \\ \langle -1/2 | \end{array} \left\| \begin{array}{cc} G_z & G_x - i G_y \\ G_x + i G_y & -G_z \end{array} \right.$$

(7.23)

Note that this is an example of a so-called Hermitian matrix, which implies that all off-diagonal elements on one side of the diagonal are the complex conjugate of those on the mirror side (i.e., $a_{ij} = a^*_{ji}$) and, as a consequence, all eigenvalues must be real. Diagonalization gives the stationary energy levels as eigenvalues, and since the matrix in this example is of dimensionality 2×2, the problem can be readily solved analytically with the determinant equation

$$\left| E - \lambda I \right| = 0$$

(7.24)

in which I is the 2×2 unit matrix, and λ are the roots of the quadratic equation, that is,

$$\begin{vmatrix} G_z - \lambda & G_x - iG_y \\ G_x + iG_y & -G_z - \lambda \end{vmatrix} = 0 \qquad (7.25)$$

or written out

$$(G_z - \lambda)(-G_z - \lambda) - (G_x + iG_y)(G_x - iG_y) = 0 \qquad (7.26)$$

which has the two solutions

$$\lambda_i \equiv E_i = \pm\sqrt{G_x^2 + G_y^2 + G_z^2} \qquad (7.27)$$

and their difference, ΔE_i, affords the resonance condition for a rhombic $S = 1/2$ system previously given in Equations 5.4 and 5.5, namely

$$h\nu = \sqrt{g_x^2 l_x^2 + g_y^2 l_y^2 + g_z^2 l_z^2}\,\beta B \qquad (7.28)$$

Matrix theory tells us that this diagonalization process can be seen as a "rotation" of the nondiagonal matrix with reference to the original basis set (Equation 7.20) to the diagonal matrix with reference to a new basis set whose wavefunctions are linear combinations of the original ones, that is,

$$\begin{pmatrix} \cos\alpha & \sin\alpha \\ -\sin\alpha & \cos\alpha \end{pmatrix} \begin{pmatrix} G_z & G_x - iG_y \\ G_x + iG_y & -G_z \end{pmatrix} \begin{pmatrix} \cos\alpha & -\sin\alpha \\ \sin\alpha & \cos\alpha \end{pmatrix} =$$

$$\begin{pmatrix} G_x + G_y + G_z & 0 \\ 0 & -G_x - G_y - G_z \end{pmatrix} \qquad (7.29)$$

where the elements of the rows of the rotation matrix on the left (or the elements of the columns in the inverse matrix on the right) are the coefficients of the original wavefunctions in the new basis set:

$$\psi_1 = \cos\alpha\left|+1/2\right\rangle + \sin\alpha\left|-1/2\right\rangle$$

$$\psi_2 = -\sin\alpha\left|+1/2\right\rangle + \cos\alpha\left|-1/2\right\rangle \qquad (7.30)$$

which, with a_{ij} being the elements of the nondiagonal matrix (e.g., $a_{12} = G_x - iG_y$), are defined by

$$\tan 2\alpha = -(a_{12} + a_{21})/(a_{11} - a_{22}) = -G_x / G_z$$

$$2\cos^2\alpha = 1 + (1 + \tan^2 2\alpha)^{-0.5} \qquad (7.31)$$

$$2\sin^2\alpha = 1 - (1 + \tan^2 2\alpha)^{-0.5}$$

These coefficients (Equation 7.30) are required to calculate the transition probability or spectral amplitude (cf. Chapter 8). Note that for systems with more than two spin wavefunctions (S > 1/2) the energy eigenvalue problem is usually not solvable analytically (unless the matrix can be reduced to one of lower dimensionality because it has sufficient off-diagonal elements equal to zero) and numerical diagonalization is the only option.

7.4   NOT-SO-SIMPLE EXAMPLE: S = 3/2

Systems with more than one unpaired electron are not only subject to the electronic Zeeman interaction but also to the magnetic-field independent interelectronic zero-field interaction, and the spin Hamiltonian then becomes

$$H_S = \beta B \bullet g \bullet S + S \bullet D \bullet S \tag{7.32}$$

in which we have now replaced the use of an asterisk to indicate unspecified interactions by the use of a dot for specified interactions: the Zeeman term encompasses multiplication of a row vector B by a matrix g by a column vector operator S, and the zero-field term is a multiplication of a row vector operator S by a matrix D by a column operator S:

$$H_S = \beta \begin{pmatrix} B_x & B_y & B_z \end{pmatrix} \begin{pmatrix} g_x & 0 & 0 \\ 0 & g_y & 0 \\ 0 & 0 & g_z \end{pmatrix} \begin{pmatrix} S_x \\ S_y \\ S_z \end{pmatrix} + \begin{pmatrix} S_x & S_y & S_z \end{pmatrix} \begin{pmatrix} D_x & 0 & 0 \\ 0 & D_y & 0 \\ 0 & 0 & D_z \end{pmatrix} \begin{pmatrix} S_x \\ S_y \\ S_z \end{pmatrix} \tag{7.33}$$

However, we have previously seen in Chapter 5 that the number of elements in the zero-field tensor can be reduced to two (Equation 5.25) by making D traceless, and so the spin Hamiltonian can be written as

$$H_S = 2G_x S_x + 2G_y S_y + 2G_z S_z + D\left[S_z^2 - S(S+1)/3 \right] + E\left(S_x^2 - S_y^2 \right) \tag{7.34}$$

The expression contains squared spin operators, which means that they have to be applied twice to the basic set of spin functions, and, for example, operator S_x^2 connects $|m_S\rangle$ states that differ by *two* units. The original set for S = 3/2 is

$$\varphi_i = \left\{ |+3/2\rangle; |+1/2\rangle; |-1/2\rangle; |-3/2\rangle \right\} \tag{7.35}$$

and the written out spin operations are

$$S_z|+3/2\rangle = (3/2)|+3/2\rangle$$

$$S_z|+1/2\rangle = (1/2)|+1/2\rangle$$

$$S_z|-1/2\rangle = -(1/2)|-1/2\rangle \tag{7.36a}$$

$$S_z|-3/2\rangle = -(3/2)|-3/2\rangle$$

$$S_z^2|+3/2\rangle = (3/2)S_z|+3/2\rangle = (9/4)|+3/2\rangle$$
$$S_z^2|+1/2\rangle = (1/2)S_z|+1/2\rangle = (1/4)|+1/2\rangle$$
$$S_z^2|-1/2\rangle = -(1/2)S_z|-1/2\rangle = (1/4)|-1/2\rangle$$
$$S_z^2|-3/2\rangle = -(3/2)S_z|-3/2\rangle = (9/4)|-3/2\rangle$$

(7.36b)

$$S_+|+1/2\rangle = \sqrt{(3/2)[(3/2)+1]-(1/2)[(1/2)+1]}|+3/2\rangle = \sqrt{3}|+3/2\rangle$$
$$S_+|-1/2\rangle = \sqrt{(3/2)[(3/2)+1]-(-1/2)[(-1/2)+1]}|+1/2\rangle = 2|+1/2\rangle$$
$$S_+|-3/2\rangle = \sqrt{(3/2)[(3/2)+1]-(-3/2)[(-3/2)+1]}|-1/2\rangle = \sqrt{3}|-1/2\rangle$$

(7.36c)

$$S_-|+3/2\rangle = \sqrt{(3/2)[(3/2+1]-(3/2)[(3/2)-1]}|+1/2\rangle = \sqrt{3}|+1/2\rangle$$
$$S_-|+1/2\rangle = \sqrt{(3/2)[(3/2+1]-(1/2)[(1/2)-1]}|-1/2\rangle = 2|-1/2\rangle$$
$$S_-|-1/2\rangle = \sqrt{(3/2)[(3/2)+1]-(-1/2)[(-1/2)-1]}|-3/2\rangle = \sqrt{3}|-3/2\rangle$$

(7.36d)

$$S_+^2|-1/2\rangle = 2S_+|+1/2\rangle = 2\sqrt{3}|+3/2\rangle$$
$$S_+^2|-3/2\rangle = \sqrt{3}S_+|-1/2\rangle = 2\sqrt{3}|+1/2\rangle$$

(7.36e)

$$S_-^2|+3/2\rangle = \sqrt{3}S_-|+1/2\rangle = 2\sqrt{3}|-1/2\rangle$$
$$S_-^2|+1/2\rangle = 2S_-|-1/2\rangle = 2\sqrt{3}|-3/2\rangle$$

(7.36f)

$$S_x|+3/2\rangle = (1/2)S_-|+3/2\rangle = (\sqrt{3}/2)|+1/2\rangle$$
$$S_x|+1/2\rangle = (1/2)[S_+|+1/2\rangle + S_-|+1/2\rangle] = (\sqrt{3}/2)|+3/2\rangle + |-1/2\rangle$$
$$S_x|-1/2\rangle = (1/2)[S_+|-1/2\rangle + S_-|-1/2\rangle] = |+1/2\rangle + (\sqrt{3}/2)|-3/2\rangle$$
$$S_x|-3/2\rangle = (1/2)S_+|-3/2\rangle = (\sqrt{3}/2)|-1/2\rangle$$

(7.36g)

$$S_y|+3/2\rangle = (i/2)S_-|+3/2\rangle = (i\sqrt{3}/2)|+1/2\rangle$$
$$S_y|+1/2\rangle = (-i/2)S_+|+1/2\rangle + (i/2)S_-|+1/2\rangle = (-i\sqrt{3}/2)|+3/2\rangle + i|-1/2\rangle$$
$$S_y|-1/2\rangle = (-i/2)S_+|-1/2\rangle + (i/2)S_-|-1/2\rangle = -i|+1/2\rangle + (i\sqrt{3}/2)|-3/2\rangle$$
$$S_y|-3/2\rangle = (-i/2)S_+|-3/2\rangle = (-i\sqrt{3}/2)|-1/2\rangle$$

(7.36h)

$$S_x^2\left|+3/2\right\rangle = (\sqrt{3}/2)S_x\left|+1/2\right\rangle = (3/2)\left|+3/2\right\rangle + (\sqrt{3}/2)\left|-1/2\right\rangle$$

$$S_x^2\left|+1/2\right\rangle = (\sqrt{3}/2)S_x\left|+3/2\right\rangle + S_x\left|-1/2\right\rangle = [(3/4)+1]\left|+1/2\right\rangle + (\sqrt{3}/2)\left|-3/2\right\rangle$$

$$S_x^2\left|-1/2\right\rangle = S_x\left|+1/2\right\rangle + (\sqrt{3}/2)S_x\left|-3/2\right\rangle = (\sqrt{3}/2)\left|+3/2\right\rangle + [(3/4)+1]\left|-1/2\right\rangle$$

$$S_x^2\left|-3/2\right\rangle = (\sqrt{3}/2)S_x\left|-1/2\right\rangle = (\sqrt{3}/2)\left|+1/2\right\rangle + (3/4)\left|-3/2\right\rangle$$

(7.36i)

$$S_y^2\left|+3/2\right\rangle = (i\sqrt{3}/2)S_y\left|+1/2\right\rangle = (3/2)\left|+3/2\right\rangle - (\sqrt{3}/2)\left|-1/2\right\rangle$$

$$S_y^2\left|+1/2\right\rangle = (-i\sqrt{3}/2)S_y\left|+3/2\right\rangle + iS_y\left|-1/2\right\rangle = [(3/4)+1]\left|+1/2\right\rangle - (\sqrt{3}/2)\left|-3/2\right\rangle$$

$$S_y^2\left|-1/2\right\rangle = -iS_y\left|+1/2\right\rangle + (i\sqrt{3}/2)S_y\left|-3/2\right\rangle = -(\sqrt{3}/2)\left|+3/2\right\rangle + [1+(3/4)]\left|-1/2\right\rangle$$

$$S_y^2\left|-3/2\right\rangle = (-i\sqrt{3}/2)S_y\left|-1/2\right\rangle = -(\sqrt{3}/2)\left|+1/2\right\rangle + (3/4)\left|-3/2\right\rangle$$

(7.36j)

$$\left(S_x^2 - S_y^2\right)\left|+3/2\right\rangle = \sqrt{3}\left|-1/2\right\rangle$$

$$\left(S_x^2 - S_y^2\right)\left|+1/2\right\rangle = \sqrt{3}\left|-3/2\right\rangle$$

$$\left(S_x^2 - S_y^2\right)\left|-1/2\right\rangle = \sqrt{3}\left|+3/2\right\rangle$$

$$\left(S_x^2 - S_y^2\right)\left|-3/2\right\rangle = \sqrt{3}\left|+1/2\right\rangle$$

(7.36k)

and with the Hamiltonian in Equation 7.34, and using for $S = 3/2$

$$D\left[S_z^2 - S(S+1)/3\right]\left|m_S\right\rangle = D\left[S_z^2\left|m_S\right\rangle - (5/4)\right]$$

(7.37)

this finally gives us the energy matrix

$$
\begin{array}{c|cccc}
\langle +3/2| & D+3G_z & \sqrt{3}G_- & \sqrt{3}E & 0 \\
\langle +1/2| & \sqrt{3}G_+ & -D+G_z & 2G_- & \sqrt{3}E \\
\langle -1/2| & \sqrt{3}E & 2G_+ & -D-G_z & \sqrt{3}G_- \\
\langle -3/2| & 0 & \sqrt{3}E & \sqrt{3}G_+ & D-3G_z
\end{array}
$$

(7.38)

in which we used the shorthand $G_+ = G_x + iG_y$ and $G_- = G_x - iG_y$ (and $G_i \equiv \beta B g_i l_i/2$). Note that once more the matrix is Hermitian ($a_{ij} = a*_{ji}$) and so all eigenvalues must be real.

It is informative to look at this matrix in zero field ($B = 0$):

$$
\begin{array}{c|cccc}
\langle +3/2| & D & 0 & \sqrt{3}E & 0 \\
\langle +1/2| & 0 & -D & 0 & \sqrt{3}E \\
\langle -1/2| & \sqrt{3}E & 0 & -D & 0 \\
\langle -3/2| & 0 & \sqrt{3}E & 0 & D
\end{array}
$$

(7.39)

For axial symmetry (i.e., $E = 0$) this is a diagonal matrix with only two eigenvalues: D and $-D$ and a zero-field splitting $\Delta E = D - -D = 2D$ as we have previously noted in

Figure 5.12. But even for systems with finite rhombicity the zero-field energy matrix can be reduced to 2×2 dimensionality

$$E = \begin{pmatrix} D & \sqrt{3}E \\ \sqrt{3}E & -D \end{pmatrix} \qquad (7.40)$$

attesting to the fact that for half-integer spin systems the spin energy levels always form degenerate Kramer's doublets in the absence of a magnetic field. Since Equation 7.40 is two dimensional, we can diagonalize it analytically by solving the determinant equation just as Equation 7.25, and the eigenvalues are

$$\lambda \equiv E_i = \pm\sqrt{D^2 + 3E^2} \qquad (7.41)$$

So the zero-field splitting can be seen to be a function of the rhombicity, and ranges from 2-D in axial symmetry to $(4/3)D\sqrt{3}$ for maximal rhombicity ($E/D = 1/3$).

As soon as the magnetic field is turned on, the matrix in Equation 7.38 applies, and there is no longer any simple analytical solution. We have discussed in Chapter 5 that in the limiting cases of "weak field" ($S*S \gg B*S$) or "strong field" ($S*S \ll B*S$) approximate analytical solutions can be deduced with the help of perturbation theory for specific cases (e.g., near axial or near rhombic symmetry; cf. Equations 5.27–5.38). However, in the general case of interaction strengths being of similar orders of magnitude ($S*S \approx S*B$) and/or when the rhombicity has an intermediate value, then there is no escape and we must diagonalize the full energy matrix in Equation 7.38 numerically in order to get the relative energy-level values of the spin manifold. The procedure to determine the resonance condition is rather different from the approach in Chapter 5 where we obtained a unique resonance-field value B_{res} directly by substitution of a microwave-frequency value and EPR-parameter values (e.g., g-values) in an analytical expression. On the contrary, here we must digitally step through every possible experimental value of the magnetic field B, diagonalize the energy matrix for that B-value, and check whether the resulting ladder of energy levels contains a pair (E_i, E_j) whose difference happens to fit the microwave quantum: $\Delta E \approx h\nu$. Fortunately, the CPUs of standard PCs have become sufficiently efficient to be able to deal with such a problem (with properly coded programs) almost in real time for realistic spin systems unless perhaps their spectra are complicated by the occurrence of extensive distributions in the spin-Hamiltonian parameters (Chapter 9).

As an example, in Figure 7.1 we give the effective g-values for an $S = 3/2$ system in X-band with either $D = 3.0$ cm^{-1} (solid lines) or $D = 0.3$ cm^{-1} (broken lines). The rhombogram with $D = 3.0$ cm^{-1} corresponds to the weak-field limit; it does not matter whether $D = 3$ cm^{-1} or 300 cm^{-1} or any other value > 3 cm^{-1}. The effective g-values are a function of the rhombicity only. And the axial case ($E = 0$) has $g^{eff} = 4, 4, 2$ for the $m_S = \pm 1/2$ doublet, and $g^{eff} = 6, 0, 0$ for the $m_S = \pm 3/2$ doublet as in the scheme of Figure 5.12. Contrarily, when $D = 0.3$ cm^{-1}, that is, in the intermediate-field case when $S*S \approx B*S$, the effective g-values strongly deviate from the standard rhombogram values: they are a function of both the value of D and of E. They cannot be looked up, but can only be calculated via energy matrix diagonalization for the specific values of D and E.

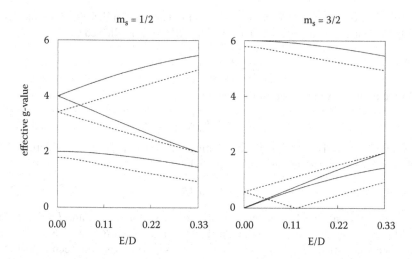

FIGURE 7.1 $S = 3/2$ rhombograms for weak-field and intermediate-field conditions. Effective g-values are plotted for $\upsilon = 9.50$ GHz and $D = 3.0$ cm^{-1} (weak field: solid lines) or $D = 0.3$ cm^{-1} (intermediate field: broken lines) as a function of the rhombicity E/D.

7.5 CHALLENGING EXAMPLE: INTEGER SPIN $S \geq 2$

How do we know or decide what terms to put in the spin Hamiltonian? This is a question of rather far-reaching importance because, since we look at our biomolecular systems through the framework of the spin Hamiltonian, our initial choice very much determines the quality limits of our final results. In other branches of spectroscopy this is sometimes referred to as a "sporting" activity. We are guided (one would hope) by a fine balance of intellectual inspection, (bio)chemical intuition, and practical considerations. In a more hypochondriacal vein, one could also call this the Achilles' heel of the spectroscopy: a wrong choice of the model (the spin Hamiltonian) will not lead to an accurate description of nature represented by the paramagnetic biomolecule.

Our starting point is the $X*Y$ interaction scheme outlined in Chapter 5 (Figure 5.2; Table 5.3). Since we have chosen to do spectroscopy on paramagnetic molecules, and since our spectrometer has a magnet, the electronic Zeeman interaction $B*S$ should always be there in our model (it is possible to do zero-field EPR spectroscopy (Bramley and Strach 1983) but I know of no biomolecular applications). However, in the theory of the spin Hamiltonian the interaction between electron spins and an external magnetic field is actually described by an—in principle, unlimited—expansion of terms. In other words, our familiar Zeeman expression $\beta B \bullet g \bullet S$ is only the leading term in an expansion that contains terms of higher order in B and/or S. For example, terms of the form B^3*S or of the form $B*S^3$ are theoretically allowed. The former is only expected to become significant at extremely high fields, but the latter might perhaps have significance for high-spin systems in high-field EPR. No biomolecular EPR observation of such terms has ever been reported yet, so why do we bother to deal with this esoteric matter at all? The reason is that at least one particular type of spin Hamiltonian term, not covered by our bilinear $X*Y$ picture, is presumably quite relevant for bioEPR.

For the sake of completeness we note that spin-Hamiltonian theory (Pake and Estle 1973: 95) states that any interaction of the form $B^b * S^s * I^i$ is allowed to occur, i.e.,

$$H_S = \sum_{b,s,i} B^b * S^s * I^i \tag{7.42}$$

with the only theoretical restrictions that

$$s \le 2S; \; i \le I; \; b+s+i = even \tag{7.43}$$

To keep the analysis tractable by eliminating insignificantly small interactions, the following practical restrictions are usually added

$$b \le 1; \; i \le 2 \tag{7.44}$$

This still leaves a considerable number of terms, e.g., for $S = 3/2$ and $I \ne 0$, one has 16 different terms in addition to the familiar five main ones $B*S$, $S*S$, $S*I$, $B*I$, and $I*I$ (Pake and Estle 1973: Table 4.1). None of these terms has ever been reported to be of significance in bioEPR except for the ones with $b = 0$ and $i = 0$, which, in our asterisk notation, are of the form $S*S*S*S$, $S*S*S*S*S*S$, etc. (or: S^4, S^6, etc.).

We now single out one of these interactions for our discussion of the integer-spin $S = 2$ system, and we defer an explanation of this deliberate choice to the end of this section. We write the spin Hamiltonian for an isolated system (i.e., no interactions between paramagnets) with $S = 2$ and $I = 0$ (i.e., no hyperfine interactions) as

$$H_S = 2G_x S_x + 2G_y S_y + 2G_z S_z + D\left[S_z^2 - S(S+1)/3\right] + E\left(S_x^2 - S_y^2\right) + a_4\left(S_+^4 + S_-^4\right) \tag{7.45}$$

which is identical to the Equation 7.34 used for $S = 3/2$ with the addition of a term in S^4. Let us once more go through the routine of developing the energy matrix.

Following our standard routine we would choose our initial orthogonal basis set to be

$$\varphi = \left\{|+2\rangle; \; |-2\rangle; \; |+1\rangle; \; |-1\rangle; \; |0\rangle\right\} \tag{7.46}$$

but since we know that any linear combination of this set will do equally well as an initial choice (i.e., the final result of a diagonalized energy matrix does not depend on the initial choice of the basis set), we could equally well start off with the set

$$\xi_1 \equiv |2^{sym}\rangle = \frac{1}{\sqrt{2}}(|+2\rangle + |-2\rangle)$$

$$\xi_2 \equiv |2^{anti}\rangle = \frac{1}{\sqrt{2}}(|+2\rangle - |-2\rangle)$$

$$\xi_3 \equiv |1^{sym}\rangle = \frac{1}{\sqrt{2}}(|+1\rangle + |-1\rangle) \tag{7.47}$$

$$\xi_4 \equiv |1^{anti}\rangle = \frac{1}{\sqrt{2}}(|+1\rangle - |-1\rangle)$$

$$\xi_5 = |0\rangle$$

which is common practice when considering integer-spin systems because when written out on this basis, the zero-field ($B = 0$) energy matrix is particularly informative, as we will see, below. First, we write out the spin operations

$$S_z\left|2^{sym}\right\rangle = (1/\sqrt{2})(S_z\left|+2\right\rangle + S_z\left|-2\right\rangle)) = (1/\sqrt{2})(2\left|+2\right\rangle - 2\left|-2\right\rangle) = 2\left|2^{anti}\right\rangle$$

$$S_z\left|2^{anti}\right\rangle = (1/\sqrt{2})(S_z\left|+2\right\rangle - S_z\left|-2\right\rangle)) = (1/\sqrt{2})(2\left|+2\right\rangle + 2\left|-2\right\rangle) = 2\left|2^{sym}\right\rangle$$

$$S_z\left|1^{sym}\right\rangle = (1/\sqrt{2})(S_z\left|+1\right\rangle + S_z\left|-1\right\rangle)) = (1/\sqrt{2})(1\left|+1\right\rangle - 1\left|-1\right\rangle) = 1\left|1^{anti}\right\rangle \qquad (7.48a)$$

$$S_z\left|1^{anti}\right\rangle = (1/\sqrt{2})(S_z\left|+1\right\rangle - S_z\left|-1\right\rangle)) = (1/\sqrt{2})(1\left|+1\right\rangle + 1\left|-1\right\rangle) = 1\left|1^{sym}\right\rangle$$

$$S_z\left|0\right\rangle = 0\left|0\right\rangle$$

$$S_z^2\left|2^{sym}\right\rangle = 2S_z\left|2^{anti}\right\rangle = 4\left|2^{sym}\right\rangle$$

$$S_z^2\left|2^{anti}\right\rangle = 2S_z\left|2^{sym}\right\rangle = 4\left|2^{anti}\right\rangle$$

$$S_z^2\left|1^{sym}\right\rangle = S_z\left|1^{anti}\right\rangle = 1\left|1^{sym}\right\rangle \qquad (7.48b)$$

$$S_z^2\left|1^{anti}\right\rangle = S_z\left|1^{sym}\right\rangle = 1\left|1^{anti}\right\rangle$$

$$S_z^2\left|0\right\rangle = 0$$

$$S_+\left|2^{sym}\right\rangle = (1/\sqrt{2})S_+\left|-2\right\rangle = (1/\sqrt{2})\sqrt{2(2+1) - -2(-2+1)}\left|-1\right\rangle = \sqrt{2}\left|-1\right\rangle$$

$$S_+\left|2^{anti}\right\rangle = -(1/\sqrt{2})S_+\left|-2\right\rangle = -\sqrt{2}\left|-1\right\rangle$$

$$S_+\left|1^{sym}\right\rangle = (1/\sqrt{2})(S_+\left|+1\right\rangle + S_+\left|-1\right\rangle) =$$

$$= (1/\sqrt{2})(\sqrt{2(2+1) - 1(1+1)}\left|+2\right\rangle + \sqrt{2(2+1) - -1(-1+1)}\left|0\right\rangle) = \sqrt{2}\left|+2\right\rangle + \sqrt{3}\left|0\right\rangle \qquad (7.48c)$$

$$S_+\left|1^{anti}\right\rangle = \sqrt{2}\left|+2\right\rangle - \sqrt{3}\left|0\right\rangle$$

$$S_+\left|0\right\rangle = \sqrt{(2(2+1) - 0(0+1)}\left|+1\right\rangle = \sqrt{6}\left|+1\right\rangle$$

$$S_-\left|2^{sym}\right\rangle = (1/\sqrt{2})S_-\left|+2\right\rangle = (1/\sqrt{2})\sqrt{2(2+1) - 2(2-1)}\left|+1\right\rangle = \sqrt{2}\left|+1\right\rangle$$

$$S_-\left|2^{anti}\right\rangle = (1/\sqrt{2})S_-\left|+2\right\rangle = \sqrt{2}\left|+1\right\rangle$$

$$S_-\left|1^{sym}\right\rangle = (1/\sqrt{2})(S_-\left|+1\right\rangle + S_-\left|-1\right\rangle) =$$

$$= (1/\sqrt{2})(\sqrt{2(2+1) - 1(1-1)}\left|0\right\rangle + \sqrt{2(2+1) - -1(-1-1)}\left|-2\right\rangle)) = \sqrt{3}\left|0\right\rangle + \sqrt{2}\left|-2\right\rangle \qquad (7.48d)$$

$$S_-\left|1^{anti}\right\rangle = (1/\sqrt{2})S_-\left|+1\right\rangle - S_-\left|-1\right\rangle = \sqrt{3}\left|0\right\rangle - \sqrt{2}\left|-2\right\rangle$$

$$S_-\left|0\right\rangle = \sqrt{2(2+1) - 0(0-1)}\left|-1\right\rangle = \sqrt{6}\left|-1\right\rangle$$

$$S_+^2 \left| 2^{sym} \right\rangle = \sqrt{2} S_+ \left| -1 \right\rangle = \sqrt{2} \sqrt{2(2+1) - -1(-1+1)} \left| 0 \right\rangle = 2\sqrt{3} \left| 0 \right\rangle$$

$$S_+^2 \left| 2^{anti} \right\rangle = -2\sqrt{3} \left| 0 \right\rangle$$

$$S_+^2 \left| 1^{sym} \right\rangle = \sqrt{3} S_+ \left| 0 \right\rangle = \sqrt{3} \sqrt{2(2+1) - 0(0+1)} \left| +1 \right\rangle = 3\sqrt{2} \left| +1 \right\rangle \qquad (7.48\text{e})$$

$$S_+^2 \left| 1^{anti} \right\rangle = -3\sqrt{2} \left| +1 \right\rangle$$

$$S_+^2 \left| 0 \right\rangle = \sqrt{6} S_+ \left| +1 \right\rangle = \sqrt{6} \sqrt{2(2+1) - 1(1+1)} \left| +2 \right\rangle = 2\sqrt{6} \left| +2 \right\rangle$$

$$S_-^2 \left| 2^{sym} \right\rangle = \sqrt{2} S_- \left| +1 \right\rangle = \sqrt{2} \sqrt{2(2+1) - 1(1-1)} \left| 0 \right\rangle = 2\sqrt{3} \left| 0 \right\rangle$$

$$S_-^2 \left| 2^{anti} \right\rangle = 2\sqrt{3} \left| 0 \right\rangle$$

$$S_-^2 \left| 1^{sym} \right\rangle = \sqrt{3} S_- \left| 0 \right\rangle = \sqrt{3} \sqrt{2(2+1) - 0(0-1)} \left| -1 \right\rangle = 3\sqrt{2} \left| -1 \right\rangle \qquad (7.48\text{f})$$

$$S_-^2 \left| 1^{anti} \right\rangle = 3\sqrt{2} \left| -1 \right\rangle$$

$$S_-^2 \left| 0 \right\rangle = \sqrt{6} S_- \left| -1 \right\rangle = \sqrt{6} \sqrt{2(2+1) - -1(-1-1)} \left| -2 \right\rangle = 2\sqrt{6} \left| -2 \right\rangle$$

$$S_+^4 \left| 2^{sym} \right\rangle = 2\sqrt{3} S_+^2 \left| 0 \right\rangle = (2\sqrt{3}) 2\sqrt{6} \left| +2 \right\rangle = 12\sqrt{2} \left| +2 \right\rangle$$

$$S_+^4 \left| 2^{anti} \right\rangle = -2\sqrt{3} S_+^2 \left| 0 \right\rangle = -12\sqrt{2} \left| +2 \right\rangle$$

$$S_-^4 \left| 2^{sym} \right\rangle = 2\sqrt{3} S_-^2 \left| 0 \right\rangle = (2\sqrt{3}) 2\sqrt{6} \left| -2 \right\rangle = 12\sqrt{2} \left| -2 \right\rangle \qquad (7.48\text{g})$$

$$S_-^4 \left| 2^{anti} \right\rangle = 2\sqrt{3} S_-^2 \left| 0 \right\rangle = 12\sqrt{2} \left| -2 \right\rangle$$

$$(S_+^4 + S_-^4) \left| 2^{sym} \right\rangle = 12\sqrt{2} \left| +2 \right\rangle + 12\sqrt{2} \left| -2 \right\rangle = 12 \left| 2^{sym} \right\rangle$$

$$(S_+^4 + S_-^4) \left| 2^{anti} \right\rangle = -12\sqrt{2} \left| +2 \right\rangle + 12\sqrt{2} \left| -2 \right\rangle = -12 \left| 2^{anti} \right\rangle \qquad (7.48\text{h})$$

$$S_x \left| 2^{sym} \right\rangle = (1/2)(S_+ \left| 2^{sym} \right\rangle + S_- \left| 2^{sym} \right\rangle) = (1/\sqrt{2})(\left| -1 \right\rangle + \left| +1 \right\rangle) = 1 \left| 1^{sym} \right\rangle$$

$$S_x \left| 2^{anti} \right\rangle = (1/2)(S_+ \left| 2^{anti} \right\rangle + S_- \left| 2^{anti} \right\rangle) = (1/\sqrt{2})(-\left| -1 \right\rangle + \left| +1 \right\rangle) = 1 \left| 1^{anti} \right\rangle$$

$$S_x \left| 1^{sym} \right\rangle = (1/2)(S_+ \left| 1^{sym} \right\rangle + S_- \left| 1^{sym} \right\rangle) =$$

$$= (1/2)(\sqrt{2} \left| +2 \right\rangle + \sqrt{3} \left| 0 \right\rangle + \sqrt{3} \left| 0 \right\rangle + \sqrt{2} \left| -2 \right\rangle) = 1 \left| 2^{sym} \right\rangle + \sqrt{3} \left| 0 \right\rangle \qquad (7.48\text{i})$$

$$S_x \left| 1^{anti} \right\rangle = (1/2)(S_+ \left| 1^{anti} \right\rangle + S_- \left| 1^{anti} \right\rangle) =$$

$$= (1/2)(\sqrt{2} \left| +2 \right\rangle - \sqrt{3} \left| 0 \right\rangle + \sqrt{3} \left| 0 \right\rangle - \sqrt{2} \left| -2 \right\rangle = 1 \left| 2^{anti} \right\rangle + 0 \left| 0 \right\rangle$$

$$S_x \left| 0 \right\rangle = (1/2)(S_+ \left| 0 \right\rangle + S_- \left| 0 \right\rangle) = (1/2)(\sqrt{6} \left| +1 \right\rangle + \sqrt{6} \left| -1 \right\rangle) = \sqrt{3} \left| 1^{sym} \right\rangle$$

$$S_y \left| 2^{sym} \right\rangle = (1/2i)(S_+ \left| 2^{sym} \right\rangle - S_- \left| 2^{sym} \right\rangle) = (1/2i)(\sqrt{2} \left| -1 \right\rangle - \sqrt{2} \left| +1 \right\rangle) = i \left| 1^{anti} \right\rangle$$

$$S_y \left| 2^{anti} \right\rangle = (1/2i)(S_+ \left| 2^{anti} \right\rangle - S_- \left| 2^{anti} \right\rangle) = (1/2i)(-\sqrt{2} \left| -1 \right\rangle - \sqrt{2} \left| +1 \right\rangle) = i \left| 1^{sym} \right\rangle$$

$$S_y \left| 1^{sym} \right\rangle = (1/2i)(S_+ \left| 1^{sym} \right\rangle - S_- \left| 1^{sym} \right\rangle) =$$

$$= (1/2i)(\sqrt{2} \left| +2 \right\rangle + \sqrt{3} \left| 0 \right\rangle - \sqrt{3} \left| 0 \right\rangle - \sqrt{2} \left| -2 \right\rangle) = -i \left| 2^{anti} \right\rangle + 0 \left| 0 \right\rangle \qquad (7.48\text{j})$$

$$S_y \left| 1^{anti} \right\rangle = (1/2i)(S_+ \left| 1^{anti} \right\rangle - S_- \left| 1^{anti} \right\rangle) =$$

$$= (1/2i)(\sqrt{2} \left| +2 \right\rangle - \sqrt{3} \left| 0 \right\rangle - \sqrt{3} \left| 0 \right\rangle + \sqrt{2} \left| -2 \right\rangle) = -i \left| 2^{sym} \right\rangle + i\sqrt{3} \left| 0 \right\rangle$$

$$S_y \left| 0 \right\rangle = (1/2i)(S_+ \left| 0 \right\rangle - S_- \left| 0 \right\rangle) = (1/2i)(\sqrt{6} \left| +1 \right\rangle - \sqrt{6} \left| -1 \right\rangle) = -i\sqrt{3} \left| 1^{anti} \right\rangle$$

$$S_x^2 \left| 2^{sym} \right\rangle = S_x \left| 1^{sym} \right\rangle = 1 \left| 2^{sym} \right\rangle + \sqrt{3} \left| 0 \right\rangle$$

$$S_x^2 \left| 2^{anti} \right\rangle = S_x \left| 1^{anti} \right\rangle = 1 \left| 2^{anti} \right\rangle$$

$$S_x^2 \left| 1^{sym} \right\rangle = S_x \left| 2^{sym} \right\rangle + \sqrt{3} S_x \left| 0 \right\rangle = 1 \left| 1^{sym} \right\rangle + 3 \left| 1^{sym} \right\rangle = 4 \left| 1^{sym} \right\rangle \qquad (7.48\text{k})$$

$$S_x^2 \left| 1^{anti} \right\rangle = S_x \left| 2^{anti} \right\rangle = 1 \left| 1^{anti} \right\rangle$$

$$S_x^2 \left| 0 \right\rangle = \sqrt{3} S_x \left| 1^{sym} \right\rangle = \sqrt{3} \left| 2^{sym} \right\rangle + 3 \left| 0 \right\rangle$$

$$S_y^2 \left| 2^{sym} \right\rangle = i S_y \left| 1^{anti} \right\rangle = 1 \left| 2^{sym} \right\rangle - \sqrt{3} \left| 0 \right\rangle$$

$$S_y^2 \left| 2^{anti} \right\rangle = i S_y \left| 1^{sym} \right\rangle = 1 \left| 2^{anti} \right\rangle$$

$$S_y^2 \left| 1^{sym} \right\rangle = -i S_y \left| 2^{anti} \right\rangle = 1 \left| 1^{sym} \right\rangle \qquad (7.48\text{l})$$

$$S_y^2 \left| 1^{anti} \right\rangle = -i S_y \left| 2^{sym} \right\rangle + i\sqrt{3} S_y \left| 0 \right\rangle = 1 \left| 1^{anti} \right\rangle + 3 \left| 1^{anti} \right\rangle = 4 \left| 1^{anti} \right\rangle$$

$$S_y^2 \left| 0 \right\rangle = -i\sqrt{3} S_y \left| 1^{anti} \right\rangle = -\sqrt{3} \left| 2^{sym} \right\rangle + 3 \left| 0 \right\rangle$$

$$(S_x^2 - S_y^2) \left| 2^{sym} \right\rangle = 2\sqrt{3} \left| 0 \right\rangle$$

$$(S_x^2 - S_y^2) \left| 2^{anti} \right\rangle = 0 \left| 2^{anti} \right\rangle$$

$$(S_x^2 - S_y^2) \left| 1^{sym} \right\rangle = 3 \left| 1^{sym} \right\rangle \qquad (7.48\text{m})$$

$$(S_x^2 - S_y^2) \left| 1^{anti} \right\rangle = -3 \left| 1^{anti} \right\rangle$$

$$(S_x^2 - S_y^2) \left| 0 \right\rangle = 2\sqrt{3} \left| 2^{sym} \right\rangle$$

We use these relations to write out the energy matrix initially in zero field (i.e., ignoring the Zeeman interaction):

$$
\begin{array}{c|ccccc}
\langle 2^{sym}| & 2D+12a_4 & 0 & 0 & 0 & 2\sqrt{3}E \\
\langle 2^{anti}| & 0 & 2D-12a_4 & 0 & 0 & 0 \\
\langle 1^{sym}| & 0 & 0 & -D+3E & 0 & 0 \\
\langle 1^{anti}| & 0 & 0 & 0 & -D-3E & 0 \\
\langle 0| & 2\sqrt{3}E & 0 & 0 & 0 & -2D
\end{array}
\qquad (7.49)
$$

In the absence of rhombicity ($E = 0$) the matrix has three eigenvalues in terms of the axial zero-field splitting, which creates two doublets and a singulet as previously described in Chapter 5 (Figure 5.16). However, the doublets in the spin manifold of integer-spin systems are *non*-Kramer's doublets (i.e., their degeneracy can be lifted even in the absence of an external magnetic field), and here a splitting in the $\{|2^{sym}\rangle; |2^{anti}\rangle\}$ doublet equal to $24a_4$ is induced by the fourth-power term in S: $a_4(S_+^4 + S_-^4)$ in the spin Hamiltonian of Equation 7.45. As a consequence, the originally (in Figure 5.15) predicted effective g-value of 8 (for $g_{real} = 2$) will shift towards higher values (lower fields). With a finite rhombicity represented by the term $E(S_x^2 - S_y^2)$, a similar first-order splitting is induced in the $\{|1^{sym}\rangle; |1^{anti}\rangle\}$ doublet of magnitude $6E$. In practice, the magnitude of the S^4 term is usually small, and the splitting $24a_4$ is frequently (much) smaller than the splitting $6E$ from the rhombic term. In fact, the latter is often found to be greater than the microwave quantum, $6E > h\nu$, and this eliminates the possibility for a transition within the doublet. For example, for an axial zero-field parameter with typical magnitude $D = 2$ cm^{-1}, a small rhombicity of $E/D > 0.025$ would make the splitting $6E > 0.3$ cm^{-1} ($h\nu = 0.3$ cm^{-1} for 9.0 GHz). In second order the E-term also adds to the splitting in the $\{|2^{sym}\rangle; |2^{anti}\rangle\}$ doublet by mixing the $|2^{sym}\rangle$ and the $|0\rangle$ states through the off-diagonal element $2\sqrt{3}E$. In zero field this can be calculated exactly by diagonalizing the 2×2 submatrix (in which we write "a" as a shorthand for a_4)

$$
\begin{array}{c|cc}
\langle 2^{sym}| & 2D+12a & 2\sqrt{3}E \\
\langle 0| & 2\sqrt{3}E & -2D
\end{array}
\qquad (7.50)
$$

which for $a = 0$ has the roots

$$
\lambda = \pm 2\sqrt{D^2 + 3E^2}
\qquad (7.51)
$$

and this gives a splitting in the $\{|2^{sym}\rangle; |2^{anti}\rangle\}$ doublet of

$$
\Delta_2 = 2\left(\sqrt{D^2 + 3E^2} - D\right)
\qquad (7.52)
$$

which is much smaller than the $\Delta_1 = 6E$ splitting of the middle doublet. In our example of $D = 2$ cm^{-1} and $E/D = 0.025$ (i.e., $\Delta_1 = 0.3$ cm^{-1}), the spitting in the upper doublet from the shift in eigenvalue of the $|2^{sym}\rangle$ state mixing with the $|0\rangle$ state given in Equation 7.52 equals $\Delta_2 = 0.0037$ cm^{-1}.

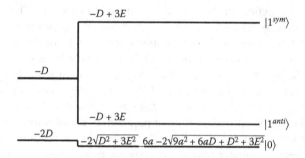

FIGURE 7.2 Zero-field manifold for $S = 2$. The energy levels on the left hand are for axial symmetry ($E = 0$ and $a = 0$), that is, two non-Kramer's doublets and a singulet. The degeneracy of the doublets is lifted by addition of an E-term and subsequent addition of an a-term.

For $a \neq 0$ the eigenvalues of Equation 7.50 are

$$\lambda = 6a \pm 2\sqrt{9a^2 + 6aD + D^2 + 3E^2} \tag{7.53}$$

and the splitting of the upper doublet becomes

$$\Delta_2 = 2\left(3a + \sqrt{9a^2 + 6aD + D^2 + 3E^2} - D\right) \tag{7.54}$$

Again, in our example of $D = 2$ cm^{-1} and $E/D = 0.025$, even a small a-value of $a = E/10$ gives $\Delta_2 = 0.064$ cm^{-1}. The complete zero-field energy manifold is schematically depicted in Figure 7.2.

To find out what the X-band spectrum of such a system will look like, let us now complete the energy matrix with the Zeeman interaction using all the spin-operations written out in Equations 7.48a to 7.48m:

$$
\begin{array}{c|ccccc}
\langle 2^{sym}| & 2D+12a_4 & 4G_z & 2G_x & i2G_y & 2\sqrt{3}E \\
\langle 2^{anti}| & 4G_z & 2D-12a_4 & i2G_y & 2G_x & 0 \\
\langle 1^{sym}| & 2G_x & -i2G_y & -D+3E & 2G_z & 2\sqrt{3}G_x \\
\langle 1^{anti}| & -i2G_y & 2G_x & 2G_z & -D-3E & i2\sqrt{3}G_y \\
\langle 0| & 2\sqrt{3}E & 0 & 2\sqrt{3}G_x & -i2\sqrt{3}G_y & -2D
\end{array}
\tag{7.55}
$$

The situation shown in Figure 7.2 is remarkable in the sense that one of the doublets of the manifold in many cases—except for systems with pronounced rhombicity (i.e., greater than $\eta \approx 0.2$)—has a relatively *small* splitting in zero field caused either by the quartic cubic term $a_4(S_+^4 + S_-^4)$ or by the rhombic quadratic term $E(S_x^2 - S_y^2)$ or both. The relevance of this observation is that the small splitting, when combined with the Zeeman interaction, causes a mixing of the original wavefunctions such that the intradoublet transition has a high probability along the z-axis for parallel-mode EPR ($B_1 \parallel B$). In regular EPR ($B_1 \perp B$) the probability is zero along z but is significant for B rotated away from z over relatively small angles. What does this all mean in practice for the X-band spectrum? In Chapter 5 (Figure 5.13), we predicted for axial symmetry and $g^{real} = 2$ an intradoublet transition with effective g_{zyx}-values of 8, 0, 0 for the highest doublet and 4, 0, 0 for the second doublet (i.e., two spectra each with zero intensity). Now rhombicity and/or a finite cubic term causes the highest doublet to split, and the feature at $g_z^{eff} = 8$ does not only get a finite amplitude in parallel-mode EPR, but it also moves to *higher* effective g_z-value: $g_z^{eff} = 8 + \delta$ where δ can take any value $0 < \delta < \infty$ depending on the magnitudes of the spin-Hamiltonian parameters a and E (or rather E/D). An infinite value for δ (and therefore for g_z^{eff}) means that the zero-field intradoublet splitting has become greater than the microwave quantum, and so a transition is no longer possible: its resonance position has, as it were, moved and disappeared into zero field. In regular, perpendicular-mode EPR the transition has only finite probability off z-axis and so a feature may be observed at an effective g-value $g^{eff} < 8 + \delta$ (which may either be greater than or less than 8). Note that the symbol g^{eff} for the off-axis feature has no subscript because it does not correspond to a molecular axis. The zero-field splitting for the $\{|1^{sym}\rangle; |1^{anti}\rangle\}$ doublet varies rapidly with the magnitude of E, and therefore we do not expect to observe any resonance at all because even if E would be small enough for the resonance not to have moved into zero field, a small distribution in E (e.g., caused by a conformational distribution) would extensively distribute the $g_z^{eff} = 4 + \delta'$ such as to broaden the spectral feature beyond detection. In summary, the predicted X-band spectrum for $S = 2$ systems in the weak-field limit ($S*S \gg B*S$) is extremely simple: a single line at $g_z^{eff} > 8$ in parallel mode and a single, weaker line at $g^{eff} < g_z^{eff}$ in perpendicular mode. Numerous X-band EPR observations on frozen solutions of $S = 2$ biomolecules and model compounds over more than 25 years are fully consistent with this prediction. A representative example was given in Figure 5.14; early examples can be found in Hagen 1982b, Hagen et al. 1984, and Hagen et al. 1985a.

It remains for us to address the question what zero-field interaction terms are allowed for a given integer spin $S = n$, and why we chose to add only the particular term $a_4(S_+^4 + S_-^4)$ to the $S = 2$ spin Hamiltonian, and then to outline how EPR intensities are computed from the wavefunctions of the diagonalized energy matrix? These issues will be addressed in the next chapter. Here, we will complete our discussion on spin Hamiltonians and energy matrices by looking at two cases in which a paramagnet does not carry one single spin but two different spins.

7.6 COMPOUNDED (OR PRODUCT) SPIN WAVEFUNCTIONS

Thus far in this chapter we have considered single-spin systems only. The zero-field interaction that we worked out in considerable detail was understood to describe interaction between unpaired electrons localized all on a single paramagnetic site with spin S and with associated spin wavefunctions defined in terms of its m_S-values, that is, $\phi = |\pm m_S\rangle$ or a linear combination of these. However, many systems of potential interest are defined by two or more different spins (cf. Figure 5.2). By means of two relatively simple examples we will now illustrate how to deal with these systems in situations where the strength of the interaction between two spins is comparable to the Zeeman interaction of at least one of them: $S_a*S_b \approx B*S_a$.

The first example concerns a system with an electron spin and a nuclear spin, and for simplicity we take $S = 1/2$ and $I = 1/2$. Actual examples would be localized radicals $^{13}C^{\bullet}$ or $^{15}N^{\bullet}$, and mononuclear low-spin $^{57}Fe^{III}$ or $^{183}W^{V}$. The spin Hamiltonian is

$$H_S = \beta B \bullet g \bullet S + S \bullet A \bullet I \qquad (7.56)$$

and assuming g and A to be colinear this can be expanded as

$$H_S = \beta B(l_x g_x S_x + l_y g_y S_y + l_z g_z S_z) + g^{-1}\left(l_x g_x^2 A_x S_x I_x + l_y g_y^2 A_y S_y S_y + l_z g_z^2 A_z S_z I_z\right) \qquad (7.57)$$

The spin wavefunctions are compounded: one part refers to the electron spin and another part to the nuclear spin, $|m_S; m_I\rangle$ (an alternative name is *product wavefunctions*):

$$\varphi(m_S, m_I) = |+1/2; +1/2\rangle; \ |+1/2; -1/2\rangle; \ |-1/2; +1/2\rangle; \ |-1/2; -1/2\rangle \qquad (7.58)$$

Just like the electron spin-raising and spin-lowering operators defined in Equations 7.15 and 7.16, we have the analogous operators for the nuclear spin I:

$$I_+\left|m_I\right\rangle = \sqrt{I(I+1) - m_I(m_I+1)}\left|m_I+1\right\rangle$$
$$I_-\left|m_I\right\rangle = \sqrt{I(I+1) - m_I(m_I-1)}\left|m_I-1\right\rangle \qquad (7.59)$$

and

$$I_x = (1/2)(I_+ + I_-)$$
$$I_y = (1/2i)(I_+ - I_-) \qquad (7.60)$$

When letting all the spin operators in the Hamiltonian of Equation 7.57 work on the compounded spin functions in Equation 7.58, note that S_i only work on the first part of the spin function, leaving the second part unchanged, and, equivalently, I_i works only on the second part leaving the first part unchanged, e.g.,

$$\left\langle +1/2; +1/2\left|S_z\right|+1/2; +1/2\right\rangle = (1/2)\left\langle +1/2; +1/2\middle|+1/2; +1/2\right\rangle = 1/2$$
$$\left\langle +1/2; +1/2\left|S_z\right|+1/2; -1/2\right\rangle = (1/2)\left\langle +1/2; +1/2\middle|+1/2; -1/2\right\rangle = 0 \qquad (7.61)$$

and for the product operators $S_i I_i$ we have e.g.,

$$S_z \left|+1/2;+1/2\right\rangle = 1/2\left|+1/2;+1/2\right\rangle$$

$$I_z \left|+1/2;+1/2\right\rangle = 1/2\left|+1/2;+1/2\right\rangle$$

$$S_z I_z \left|+1/2;+1/2\right)\rangle = (1/2)(1/2)\left|+1/2;+1/2\right\rangle = 1/4\left|+1/2;+1/2\right\rangle \tag{7.62}$$

$$\left\langle +1/2;+1/2\left|A_z S_z I_z\right|+1/2;+1/2\right\rangle = A_z/4$$

but

$$\left\langle +1/2;+1/2\left|A_z S_z I_z\right|+1/2;-1/2\right\rangle = (A_z/4)\left\langle +1/2;+1/2\right\|+1/2;-1/2\right\rangle = 0 \tag{7.63}$$

Also,

$$S_x I_x = (S^+ + S^-)(I^+ + I^-)/4 \tag{7.64}$$

and so

$$S_x I_x \left|-1/2;+1/2\right\rangle = (1/2)(1/2)\left|+1/2;-1/2\right\rangle$$

$$\left\langle +1/2;-1/2\left|A_x S_x I_x\right|-1/2;+1/2\right\rangle = A_x/4 \tag{7.65}$$

etcetera.

With all operations worked out, and using the definitions $G_i = \beta B l_i g_i/2$ and $\alpha_i = l_i g_i^2 A_i/g$, we obtain the full energy matrix

$$
\begin{array}{c|cccc}
\langle +1/2;+1/2| & G_z + \alpha_z/4 & 0 & G_x - iG_y & 0 \\
\langle +1/2;-1/2| & 0 & G_z - \alpha_z/4 & (\alpha_x + \alpha_y)/4 & G_x - iG_y \\
\langle -1/2;+1/2| & G_x + iG_y & (\alpha_x + \alpha_y)/4 & -G_z - \alpha_z/4 & 0 \\
\langle -1/2;-1/2| & 0 & G_x + iG_y & 0 & -G_z + \alpha_z/4 \\
\end{array}
\tag{7.66}
$$

which like any other Hermitian matrix can be numerically diagonalized to get the energy levels to be used in $h\nu = \Delta E$, and to obtain the coefficients of the diagonalizing basis set for the determination of intensities (cf. Chapter 8).

For biomolecular $S = 1/2$ systems subject to central hyperfine interaction the intermediate-field situation ($B*S \approx S*I$) is not likely to occur unless the microwave frequency is lowered to L-band values. When $\nu = 1$ GHz, the resonance field for $g = 2$ is at $B = 357$ gauss. Some Cu(II) sites in proteins have $A_z \approx 200$ gauss, and this would certainly define L-band EPR as a situation in which the electronic Zeeman interaction is comparable in strength to that of the copper hyperfine interaction. No relevant literature appears to be available on the subject. An early measurement of the Cu^{II}(H$_2$O)$_6$ reference system (cf. Figure 3.4) in L-band, and its simulation using the axial form of Equation 5.18 indicated that for this system

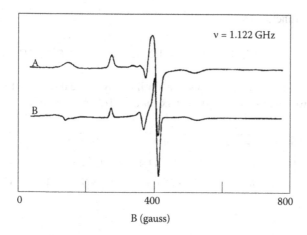

$v = 1.122$ GHz

A

B

0   400   800

B (gauss)

FIGURE 7.3 Breakdown of perturbation-theory approach for $Cu^{II}(H_2O)_6$ in L-band. The spectrum of the elongated CuO_6 octahedron (upper trace) is simulated (lower trace) with the approximative resonance condition defined in Equation 5.18. There is no fit of the first hyperfine line at low field (Hagen 1982a).

with $A_z = 131$ gauss, the perturbation-theory approximation breaks down only for the first hyperfine line at very low field as shown in Figure 7.3. Later, others have analyzed L-band data of "blue-copper" proteins using diagonalization of the proper extension of Equation 7.66 for the copper nucleus $I = 3/2$ (Antholine 1993), but the explored system does not really define an intermediate-field situation ($A_z \approx 56$ gauss; $v = 1.64$ GHz for $g = 2$ means $B_{res} \approx 586$ gauss), and the question whether this numerically demanding approach was actually necessary, was not addressed.

It would thus appear that this spectral analysis by energy matrix diagonalization of systems defined by compounded spin functions is a rather esoteric subject that we can generally ignore except when working at very low microwave frequencies. There is, however, a system that is formally quite similar to the one just treated, and which is much more likely to occur under intermediate-field conditions, even in X-band. This is the system of interacting electron spins formed by two (or more) *different* paramagnets (in contrast to the interaction between electrons of a single site). A relatively simple and very common example is the reduced 8Fe ferredoxin harboring two [4Fe-4S]$^{1+}$ cubanes within a small protein at a cube edge-to-edge distance of circa 5–10 Å.

We now consider the relatively simple case of two different electron spins each with a spin of one-half: $S_a = 1/2$ and $S_b = 1/2$. The spin Hamiltonian is

$$H_S = \beta B \bullet g_a \bullet S_a + \beta B \bullet g_b \bullet S_b + S_a \bullet D \bullet S_b \qquad (7.67)$$

and under the assumption of tensor colinearity

$$H_S = \beta B \left[l_x \left(g_{x_a} S_{x_a} + g_{x_b} S_{x_b} \right) + l_y \left(g_{y_a} S_{y_a} + g_{y_b} S_{y_b} \right) + l_z \left(g_{z_a} S_{z_a} + g_{z_b} S_{z_b} \right) \right]$$
$$+ D_x S_{x_a} S_{x_b} + D_y S_{y_a} S_{y_b} + D_z S_{z_a} S_{z_b} \qquad (7.68)$$

The spin wavefunctions are

$$\varphi_i = \left| m_{S_a} ; m_{S_b} \right\rangle = \left\{ \left| +1/2;+1/2 \right\rangle; \; \left| +1/2;-1/2 \right\rangle; \; \left| -1/2;+1/2 \right\rangle; \; \left| -1/2;-12 \right\rangle \right\} \quad (7.69)$$

and working out the spin operations in a similar way as we did above for the $S = 1/2$; $I = 1/2$ case, we obtain the energy matrix for $S_a = 1/2$ and $S_b = 1/2$:

$\langle +1/2;+1/2 \vert$	$G_z^a + G_z^b + D_z 4$	$G_x^b - iG_y^b$	$G_x^a - iG_y^a$	0
$\langle +1/2;-1/2 \vert$	$G_x^b + iG_y^b$	$G_z^a - G_z^b - D_z 4$	$(D_x + D_y)/4$	$G_x^a - iG_y^a$
$\langle -1/2;+1/2 \vert$	$G_x^a + iG_y^a$	$(D_x + D_y)/4$	$-G_z^a + G_z^b - D_z 4$	$G_x^b - iG_y^b$
$\langle -1/2;-1/2 \vert$	0	$G_x^a + iG_y^a$	$G_x^b + iG_y^b$	$-G_z^a - G_z^b + D_z 4$

$$(7.70)$$

Note that this is a very simple case indeed, because there is no reason why g_a, g_b, and D should be diagonal in the same axes system, and, furthermore, S_a and/or S_b can well be high spin.

We will return to the problem of interacting spins in Chapter 11. Here, we close with an illustration in Figure 7.4 of the typical spectral pattern from two dipolarly interacting spins in the very common example of two $S = 1/2$ [4Fe-4S]$^{1+}$ cubane clusters in a small (typically 9 kDa) ferredoxin or in a ferredoxin-like domain in a larger enzyme. While an isolated (i.e., not interacting) $S = 1/2$ cubane would give a well-defined rhombic spectrum, in the spectrum of Figure 7.4 this pattern is not easily recognized due to extra peaks and asymmetries leading to a complex spectrum with very long tails towards both low and high field (Pierik et al. 1992b).

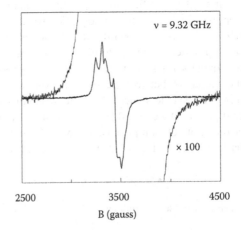

FIGURE 7.4 Example of a spin–spin interaction spectrum. The complex spectrum is from two adjacent cubane clusters in the enzyme FeFe-hydrogenase from *Desulfovibrio vulgaris*. The 100× blowup is to show the extended spectral field range resulting from interaction.

8 Biological Spin Hamiltonians

In this chapter we continue our journey into the quantum mechanics of paramagnetic molecules, while increasing our focus on aspects of relevance to biological systems. For each and every system of whatever complexity and symmetry (or the lack of it) we can, in principle, write out the appropriate spin Hamiltonian and the associated (simple or compounded) spin wavefunctions. Subsequently, we can always deduce the full energy matrix, and we can numerically diagonalize this matrix to obtain the stable energy levels of the system (and therefore all the resonance conditions), and also the coefficients of the new basis set (linear combinations of the original spin wavefunctions), which in turn can be used to calculate the transition probability, and thus the EPR amplitude of all transitions.

8.1 HIGHER POWERS OF SPIN OPERATORS

The concept of symmetry is used very frequently in physics, and somewhat less frequently in chemistry, not only because it can be aesthetically appealing to the human mind, but more so because it has the potential to simplify the complexity of problems (and the CPU time required for their numerical analysis). Physics is sometimes referred to by its protagonists as "the mother of all sciences." From the viewpoint of applicability of symmetry concepts, then, chemistry can perhaps be considered an occasionally demanding child of this mother, but molecular biology is an absolutely unruly mongrel. Not only has it long ago decided to avoid symmetry and predominantly go for L-amino acids and D-sugars, but it also employs very large and flexible ligands for its transition-ion complexes, with the effect that its coordination chemistry essentially lacks any symmetry. We are thus faced with the following dilemma: EPR theory was written by physicists (at least for the first 30 years after the discovery of the EPR effect), and relies heavily on symmetry concepts, however, we would want to apply EPR theory to biological systems without symmetry. This clash of cultures has produced what one could perhaps call "artificially enforced symmetry": the all-pervading tendency to imagine symmetries in systems that are asymmetric by nature in order to reduce problems of insurmountable complexity to tractable ones. Our use of the $a_4(S_+^4 + S_-^4)$ in the $S = 2$ spin Hamiltonian is a case in point.

The origin of zero-field terms in the spin Hamiltonian is rooted in crystal-field theory, in which coordination complexes are represented as geometric structures of point charges (Stevens 1997). The crystal-field potential of these point charges is developed in a series of homogeneous polynomials directly reflecting the symmetry of the structure. Spin Hamiltonian zero-field terms have the form of the polynomials of crystal-field theory. Thus, the symmetry of the coordination complex (be it real or artificially enforced) determines which, and how many, zero-field terms appear in the spin Hamiltonian. For example, for an $S = 2$ system of cubic symmetry (i.e., a transition metal ion at the center of a perfect octahedron of six identical ligands as in Figure 5.15A, or at the center of a perfect tetrahedron of four identical ligands) theory dictates the zero-field spin Hamiltonian to be

$$H_{cubic}^{zero} = B_4^0 O_4^0 + 5B_4^4 O_4^4 \tag{8.1}$$

in which the B_k^q's are zero-field parameters (we have thus far used D, E, and a_4) and the O_k^q's are spin operators (for example: $S_+^4 + S_-^4$). Lowering the symmetry increases the number of allowed terms. For example pulling on the two axial ligands along the z-axis lowers the symmetry to tetragonal; also, pulling on the ligand pairs in the x-y plane lowers the symmetry to orthorhombic, and finally giving the whole structure a kick from an arbitrary direction lowers the symmetry to triclinic (similar effects can be obtained through ligand substitution by nonidentical ligands). For the zero-field Hamiltonian this implies (Stevens 1952; Morin and Bonnin 1999).

$$
\begin{aligned}
H_{tetragonal}^{zero} &= B_2^0 O_2^0 + B_4^0 O_4^0 + B_4^4 O_4^4 \\[4pt]
H_{rhombic}^{zero} &= B_2^0 O_2^0 + B_2^2 O_2^2 + B_4^0 O_4^0 + B_4^2 O_4^2 + B_4^4 O_4^4 \\[4pt]
H_{triclinic}^{zero} &= B_2^0 O_2^0 + B_2^2 O_2^2 + B_4^0 O_4^0 + B_4^1 O_4^1 + B_4^{-1} O_4^{-1} \\[4pt]
&\quad + B_4^2 O_4^2 + B_4^{-2} O_4^{-2} + B_4^3 O_4^3 + B_4^{-3} O_4^{-3} + B_4^4 O_4^4 + B_4^{-4} O_4^{-4}
\end{aligned}
\tag{8.2}
$$

that is, a rapidly increasing number of terms with "decreasing" symmetry. Some of the terms in Equations 8.1 and 8.2 are familiar to us, e.g.,

$$
\begin{aligned}
O_2^0 &= 3S_z^2 - S(S+1) \\[4pt]
O_2^2 &= (1/2)\left(S_+^2 + S_-^2\right) = S_x^2 - S_y^2 \\[4pt]
O_4^4 &= (1/2)\left(S_+^4 + S_-^4\right)
\end{aligned}
\tag{8.3}
$$

which means that (cf. our initial $S = 2$ spin Hamiltonian in Equation 7.45)

$$D = B_2^0; \quad E = B_2^2; \quad a_4 = (1/2)B_4^4 \tag{8.4}$$

but all the other terms are new, for example,

$$O_4^0 = 35S_z^4 - (30S(S+1) - 25)S_z^2 - 6S(S+1) + 3S^2(S+1)^2 \tag{8.5}$$

and the remaining terms can be found in, e.g., Newman and Urban 1975. The key point to realize is that powder spectra of high-spin biomolecular systems and their models in the weak-field limit of $B*S \ll S*S$ (a common condition in X-band) exhibit only a limited number of spectral features, and this is especially true for the integer-spin systems (cf. Figure 5.14). Combined with the notion that symmetry is not a common commodity in biomolecular coordination complexes, it would appear that the analysis of their EPR spectra is a highly underdetermined problem. Chances to find a unique solution to an extended Hamiltonian as, e.g., the triclinic-H in Equation 8.2, are further complicated by the fact that most of the extra terms have spectral effects that are very similar to those of the "basic" Hamiltonian in Equation 7.45. For example, for $S = 2$ the O_4^0 operator in Equation 8.5 causes a slight shift in the non-Kramer's doublets that can be readily (mis)taken for a small modulation of the D-value of the familiar $3S_z^2 - S(S+1)$ term. All terms in Equation 8.2 are also valid for $S = 5/2$, and the same problem of underdetermination also occurs for half-integer spin systems. There is, however, one clear exception to this convolution of terms, and that is the $a_4(S_+^4 + S_-^4)$ term for $integer$-spin systems: it causes a unique first-order splitting from cubic symmetry onwards in the $\{|2^{sym}\rangle;$ $|2^{anti}\rangle\}$ doublet that cannot be mimicked by any of the other terms, and, moreover, the spectral effect of this small splitting is pronounced, as we have seen in the previous section. This is the reason why the a_4 term was added to the spin Hamiltonian for $S = 2$ in Equation 7.45.

For higher integer spins the number of allowed zero-field interaction terms further increases, and so does the convolution of comparable effects, except once more for a unique term that directly splits the highest non-Kramer's doublet. For $S \geq 3$ we have the addition, valid in cubic (and, therefore, in tetragonal, rhombic, and triclinic) symmetry:

$$H_S = B_6^6 O_6^6 = (1/2)B_6^6 \left(S_+^6 + S_-^6\right) \tag{8.6}$$

which we choose to write in our alternative notation as

$$H_{S=3} = a_6 \left(S_+^6 + S_-^6\right) \tag{8.7}$$

and for $S \geq 4$ yet another extra term appears

$$H_{S=4} = a_8 \left(S_+^8 + S_-^8\right) \tag{8.8}$$

and so on for $S \geq 5$.

As an illustration consider then a zero-field Hamiltonian for $S = 4$ in which we have retained only the familiar axial D- and rhombic E-term plus the cubic terms that split the non-Kramer's doublets in first order:

$$H_{S=4}^{zero} = D\left[S_z^2 - S(S+1)/3\right] + E\left(S_x^2 - S_y^2\right) + a_4\left(S_+^4 + S_-^4\right) + a_6\left(S_+^6 + S_-^6\right) + a_8\left(S_+^8 + S_-^8\right)$$

$$\tag{8.9}$$

which affords the zero-field energy matrix

$\langle 4^{sym}\|$ $28D/3 + 40320a_8$	0	0	0	$2\sqrt{7}E + 720\sqrt{7}a_6$	0	0	0	$24\sqrt{35}a_4$
$\langle 4^{anti}\|$ 0	$28D/3 - 40320a_8$	0	0	0	$2\sqrt{7}E - 720\sqrt{7}a_6$	0	0	0
$\langle 3^{sym}\|$ 0	0	$7D/3 + 2520a_6$	0	0	0	$3\sqrt{7}E + 60\sqrt{7}a_4$	0	0
$\langle 3^{anti}\|$ 0	0	0	$7D/3 - 2520a_6$	0	0	0	$3\sqrt{7}E - 60\sqrt{7}a_4$	0
$\langle 2^{sym}\|$ $2\sqrt{7}E + 720\sqrt{7}a_6$	0	0	0	$-8D/3 +180a_4$	0	0	0	$6\sqrt{5}E$
$\langle 2^{anti}\|$ 0	$2\sqrt{7}E - 720\sqrt{7}a_6$	0	0	0	$-8D/3 +180a_4$	0	0	0
$\langle 1^{sym}\|$ 0	0	$3\sqrt{7}E + 60\sqrt{7}a_4$	0	0	0	$-17D/3 +10E$	0	0
$\langle 1^{anti}\|$ 0	0	0	$3\sqrt{7}E - 60\sqrt{7}a_4$	0	0	0	$-17D/3 -10E$	0
$\langle 0\|$ $24\sqrt{35}a_4$	0	0	0	$6\sqrt{5}E$	0	0	0	$-20D/3$

$$(8.10)$$

Note the characteristic band structure of these energy matrices. When we first concentrate on the leading D-term in the Hamiltonian, it is seen that this axial zero-field interaction creates a manifold of one singlet and three non-Kramer's doublets separated by D, $3D$, $5D$, and $7D$, respectively, as previously depicted in Figure 5.16. The cubic a_4-term splits the $\{|2^{sym}\rangle; |2^{anti}\rangle\}$ doublet; the cubic a_6-term splits the $\{|3^{sym}\rangle; |3^{anti}\rangle\}$ doublet; and the cubic a_8-term splits the $\{|4^{sym}\rangle; |4^{anti}\rangle\}$ doublet. In lower symmetry the rhombic E-term will also split the $\{|1^{sym}\rangle; |1^{anti}\rangle\}$ doublet. Finally, the off-diagonal E-, a_4-, and a_6-terms will induce additional minor shifts in all the levels of the manifold. Comparison with the zero-field matrix for $S = 2$ in Equation 7.49 reveals that the corresponding first-order splittings are greater for $S = 4$. For example, the rhombic splitting of the $\{|1^{sym}\rangle; |1^{anti}\rangle\}$ doublet goes from $6E$ to $20E$ and the cubic splitting of the $\{|2^{sym}\rangle; |2^{anti}\rangle\}$ doublet goes from $24a_4$ to $360a_4$. In other words, if the parameters (E, a_4, etc.) are of comparable magnitude, then the likelihood to observe the intradoublet transitions $decreases$ with $increasing$ integer spin, either because the resonances disappear into zero field and/or because of increased broadening through a distribution in the zero-field parameters. However, the absolute magnitude of the a_i coefficients (or the alternative B_i^j coefficients) will presumably decrease with increasing power of the S_i operator (or, alternatively, the O_i^j operator), and so our gut feeling tells us that the highest doublet always has the smallest zero-field intradoublet splitting, and therefore, has the highest likelihood of producing an observable intradoublet transition. Lo and behold this is what we observe experimentally. Figure 8.1 shows the parallel-mode spectra of an $S = 1$, $S = 2$, $S = 3$, and an $S = 4$ system: only the transition within the highest doublet is observed, and the spectrum is presumably dominated by a single spin Hamiltonian parameter only, namely the cubic a_i parameter (or the E-parameter for S = 1). Moreover, the broadening presumably due to distributed zero-field parameters clearly decreases with increasing integer spin. Thus, the intimidating $S = 4$ energy matrix in Equation 8.10 eventually affords a single-line X-band spectrum, and inspection of that matrix clarifies why this is the case. Also, although the (bio)chemical systems that give rise to the spectra in Figure 8.1 are not likely to have any symmetry at all at their paramagnetic coordination sites, artificially enforcing initially high and subsequently lower symmetries allows us to make use of the symmetry-based tools developed by solid-state physicists for simple crystalline host materials.

The weak-field condition of $S*B \ll S*S$ is a very common one for biological systems in X-band, although it does not, of course, have generality, and a particular counter-example is the $S = 5/2$ system in certain manganese enzymes with 5-coordinated Mn^{II} (Smoukov et al. 2002; see also Chapter 12 for an example). For such systems ($S*B \approx S*S$) spectral interpretation requires numerical diagonalization of full energy matrices such as the one in Equation 8.10 (after insertion of the Zeeman terms) for $S = 4$. Furthermore, increasing the microwave frequency and the magnetic field will eventually lead to breakdown of the weak-field condition for any system. However, for many biological systems, the frequencies required for this

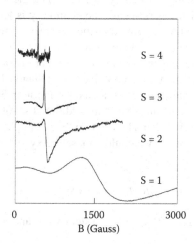

FIGURE 8.1 Intradoublet transitions for $S = 1, 2, 3, 4$. The figure shows parallel-mode spectra of the transition within the highest doublet of (from top to bottom) the $S = 4$ [8Fe-7S] P-cluster in *Xanthobacter autotrophicus* MoFe-nitrogenase, the $S = 3$ [8Fe-7S] P-cluster in *Azotobacter vinelandii* MoFe-nitrogenase, the $S = 2$ mononuclear $Fe_A(II)$ site in *Desulfovibrio vulgaris* desulfoferrodoxin, and the $S = 1$ Ni(II)EDTA complex (Hagen 2006). (Reproduced by permission of The Royal Society of Chemistry.)

breakdown correspond to values (typically > 100 GHz) at which spectrometers often have insufficient concentration sensitivity for practical applications. The literature contains numerous examples of inorganic systems (that are *not* models for biology) whose *X-band* EPR exhibits many lines as a reflection of the $S*B \approx S*S$ condition, for example, Al_2O_3 powder with dopants $S = 3/2$ Cr^{III}, $S = 5/2$ Fe^{III}, or $S = 7/2$ Gd^{III} (Priem et al. 2001).

8.2   TENSOR NONCOLINEARITY

In Equation 7.33 we have written out both the *g*-value and the zero-field coefficient of the basic S^2 interaction term in the form of diagonal 3 × 3 matrices in which all off-diagonal elements are equal to zero. The diagonal elements were indexed with subscripts x, y, z, corresponding to the Cartesian axes of the molecular axes system. But how do we define a molecular axis system in a (bio)coordination complex that lacks symmetry? The answer is that if we would have made a "wrong" choice, then the matrices would not be diagonal with zeros elsewhere. In other words, if the spin Hamiltonian would have been written out for a different axes system, then, for example, the *g*-matrix would not have three, but rather six, independent elements:

$$g = \begin{pmatrix} g_{xx} & g_{xy} & g_{xz} \\ -g_{xy} & g_{yy} & g_{yz} \\ -g_{xz} & -g_{yz} & g_{zz} \end{pmatrix} \qquad (8.11)$$

There is, in fact, only one unique axes system for which the *g*-matrix is diagonal: the molecular axes system. However, in case the symmetry is lower than orthorhombic (e.g., in the total absence of any symmetry), then the axes system in which *g* is diagonal is not necessarily identical to the axes system in which the quadratic zero-field interaction parameter *D* is diagonal, and so on for other interactions. This means that if we write out the Hamiltonian in an axes system that diagonalizes one matrix, diagonalization of a different matrix required an axes rotation in 3-D space; definition of the spectrum requires the specification of an additional set of parameters describing this interconversion of axes systems by rotation. For example, for the Hamiltonian in Equation 7.32 in low symmetry we could choose the xyz-axes system that diagonalizes *D*, but then we must write

$$H_s = \beta B \bullet R^{-1} \bullet g \bullet R \bullet S + S \bullet D \bullet S \qquad (8.12)$$

in which *g* is diagonal in a different axes system x′y′z′ and *R* is a rotation matrix in 3-D space, for example,

$$R = \begin{pmatrix} \cos\alpha\cos\beta\cos\gamma - \sin\alpha\sin\gamma & \sin\alpha\cos\beta\cos\gamma + \cos\alpha\sin\gamma & -\sin\beta\cos\gamma \\ -\cos\alpha\cos\beta\sin\gamma - \sin\alpha\cos\gamma & -\sin\alpha\cos\beta\sin\gamma + \cos\alpha\cos\gamma & \sin\beta\sin\gamma \\ \cos\alpha\sin\beta & \sin\alpha\sin\beta & \cos\beta \end{pmatrix}$$

$$\qquad (8.13)$$

This matrix describes the transformation from x'y'z' to xyz as a rotation about the z' axis over angle α, followed by a rotation about the new y" axis over angle β, followed by a final rotation over the new z"' axis over angle γ (Watanabe 1966: 148). Formally, the low-symmetry situation is even a bit more complicated because the nondiagonal g-matrix in Equation 8.11 is not necessarily skew symmetric ($g_{ij} \neq -g_{ji}$). Only the square $g \times g$ is symmetric and can be transformed into diagonal form by rotation. In mathematical terms, $g \times g$ is a second-rank tensor, and g is not.

8.3   GENERAL EPR INTENSITY EXPRESSION

In Chapter 6 we presented an expression for the transition probability (or intensity, amplitude) of field-swept spectra from randomly oriented simple $S = 1/2$ systems (Equation 6.4), and we could perhaps tacitly assume (as is generally done in the bioEPR literature) that the expression also holds for effective $S = 1/2$ systems, such as for the high-spin subspectra defined by the rhombograms discussed in Chapter 5. But what about parallel-mode spectra? And how do we compute intensities in complex situations like for systems in the $B*S \approx B*B$ intermediate-field regime? Clearly, we need a more generic approach towards intensity calculations.

We have seen that a spin Hamiltonian in combination with its associated spin wavefunctions defines an energy matrix, which can always be diagonalized to obtain all the real energy sublevels of the spin manifold. Furthermore, the diagonalization also affords a new set of spin wavefunctions that are a basis for the diagonal matrix, and which are linear combinations of the initial set of spin functions. The coefficients in these linear combinations can be used to calculate the transition probabilities of all transitions within the spin manifold.

In this section we will derive two general expressions for the transition probability: one for regular perpendicular-mode EPR ($B_1 \perp B$) and one for parallel-mode EPR ($B_1 \parallel B$). The two expressions are related in the sense that they also provide the correct ratio of intensities (perpendicular over parallel) for data obtained with a single, dual mode resonator. The expressions are derived here, and not just given, because all expressions thus far published in the EPR literature contain small inconsistencies and/or errors.

Our starting point is Fermi's golden rule for transition to a single state

$$P_{pq}(t) = (2\pi / \hbar) \left| \langle p | H_1(t) | q \rangle \right|^2 \tag{8.14}$$

in which P is the transition probability for a transition between two spin-manifold sublevels $|p\rangle$ and $|q\rangle$. The perturbing microwave Hamiltonian causing the transitions is, in the Cartesian molecular axis systems i = x, y, z,

$$H_1(t) = \beta B_1 \left(\sum_i k_i g_i S_i \right) \cos \omega t \tag{8.15}$$

in which B_1 is the magnetic-field component of the microwave, and $\cos \omega t$ is its time dependence.

The k_i's are the direction cosines of B_1 with respect to the xyz-axes system. Equations 8.14 and 8.15 combine into an expression with a system-dependent prefactor and a part that depends on the specific spin system under consideration,

$$P_{pq} = [(2\pi / \hbar)\beta B_1 \cos \omega t] \left| \sum_i k_i g_i \left\langle p|S_i|q \right\rangle \right|^2 \tag{8.16}$$

When comparing relative intensity data obtained in a single frequency band, we are not interested in the magnitude of the prefactor, therefore, we define a relative transition probability

$$W_{pq} = \sum_{m,n} k_m g_m S_m^*(p,q) k_n g_n S_n(p,q) \tag{8.17}$$

in which $\{m, n\}$ run through $\{x, y, z\}$, and

$$S_i^*(p,q) = \left\langle \sum_p p(\theta,\varphi) \middle| S_i \middle| \sum_q q(\theta,\varphi) \right\rangle \tag{8.18}$$

The spin wavefunctions $|p\rangle$ and $|q\rangle$ are those obtained after diagonalization of the complete energy matrix.

In powder EPR simulators we use the orientation of the static-field vector B with respect to the molecular xyz-axes system as the definition of molecular orientation. The orientation is defined in terms of the polar angles $\{\theta, \varphi\}$, or equivalently in terms of the direction cosines l_i, as defined previously in Equation 5.3. To solve Equation 8.17 we have to define the direction cosines k_i of B_1 in terms of the direction cosines l_i of B.

A mathematically trivial, but experimentally important case is that of parallel-mode EPR, in which $B_1 \parallel B$ and, therefore, $k_i = l_i$. Substituting the expressions for the direction cosines of Equation 5.3 into Equation 8.17 gives the parallel-mode relative intensity

$$
\begin{aligned}
W_\parallel(p,q) = \ &\sin^2\theta\cos^2\varphi g_x^2 S_x^*(pq)S_x(pq) \\
&+\sin^2\theta\sin\varphi\cos\varphi g_x g_y[S_x^*(pq)S_y(pq)+S_y^*(pq)S_x(pq)] \\
&+\cos\theta\sin\theta\cos\varphi g_x g_z[S_x^*(pq)S_z(pq)+S_z^*(pq)S_x(pq)] \\
&+\sin^2\theta\sin^2\varphi g_y^2 S_y^*(pq)S_y(pq) \\
&+\cos\theta\sin\theta\sin\varphi g_y g_z[S_y^*(pq)S_z(pq)+S_z^*(pq)S_y(pq)] \\
&+\cos^2\theta g_z^2 S_z^*(pq)S_z(pq)
\end{aligned}
\tag{8.19}
$$

For powder EPR in perpendicular mode B_1 lies in a plane perpendicular to B, and B_1 samples all possible orientations in this plane. We define a vector v in this plane by means of an angle α between B_1 and v. Then, the direction cosines of B_1 are

$$k_x = \cos\alpha\cos\theta\cos\varphi - \sin\alpha\sin\varphi$$

$$k_y = \cos\alpha\cos\theta\sin\varphi + \sin\alpha\cos\varphi$$

$$k_z = -\cos\alpha\sin\theta \tag{8.20}$$

Substituting these in Equation 8.17 gives

$$
\begin{aligned}
W_\perp(p,q) = \; & k_x^2 g_x^2 S_x^*(pq)S_x(pq) + k_x k_y g_x g_y [S_x^*(pq)S_y(pq) + S_y^*(pq)S_x(pq)] \\
& + k_y^2 g_y^2 S_y^*(pq)S_y(pq) + k_y k_z g_y g_z [S_y^*(pq)S_z(pq) + S_z^*(pq)S_y(pq)] \\
& + k_z^2 g_z^2 S_z^*(pq)S_z(pq) + k_x k_z g_x g_z [S_x^*(pq)S_z(pq) + S_z^*(pq)S_x(pq)]
\end{aligned}
\tag{8.21}
$$

in which

$$k_x^2 = \underline{\cos^2\alpha}\cos^2\theta\cos^2\varphi - 2\underline{\sin\alpha\cos\alpha}\cos\theta\sin\varphi\cos\varphi + \underline{\sin^2\alpha}\sin^2\varphi$$

$$k_y^2 = \underline{\cos^2\alpha}\cos^2\theta\sin^2\varphi - 2\underline{\sin\alpha\cos\alpha}\cos\theta\sin\varphi\cos\varphi + \underline{\sin^2\alpha}\cos^2\varphi$$

$$k_z^2 = \underline{\cos^2\alpha}\sin^2\theta$$

$$k_x k_y = \underline{\cos^2\alpha}\cos^2\theta\cos\varphi\sin\varphi + \underline{\sin\alpha\cos\alpha}\cos\theta(\cos^2\varphi - \sin^2\varphi)$$

$$\qquad\qquad - \underline{\sin^2\alpha}\sin\varphi\cos\varphi$$

$$k_x k_z = -\underline{\cos^2\alpha}\sin\theta\cos\theta\cos\varphi + \underline{\sin\alpha\cos\alpha}\sin\theta\sin\varphi$$

$$k_y k_z = -\underline{\cos^2\alpha}\sin\theta\cos\theta\sin\varphi - \underline{\sin\alpha\cos\alpha}\sin\theta\cos\varphi \tag{8.22}$$

From the underscored factors in Equation 8.22 it can be seen that there are only three types of terms in α and, therefore, that Equation 8.21 can be written as

$$W_\perp(p,q) = A\cos^2\alpha + B\sin\alpha\cos\alpha + C\sin^2\alpha \tag{8.23}$$

We integrate α in the x-y plane between the limits $0\text{--}2\pi$ to obtain

$$\frac{1}{2\pi}\int_0^{2\pi} W_\perp(p,q)d\alpha = \frac{1}{2}(A+C) \tag{8.24}$$

and writing out A and C gives the perpendicular-mode relative intensity

$$
\begin{aligned}
W_\perp(p,q) = (1/2)\{ & (1 - \sin^2\theta\cos^2\varphi)g_x^2 S_x^*(pq)S_x(pq) \\
& - \sin^2\theta\sin\varphi\cos\varphi\, g_x g_y [S_x^*(pq)S_y(pq) + S_y^*(pq)S_x(pq)] \\
& - \cos\theta\sin\theta\cos\varphi\, g_x g_z [S_x^*(pq)S_z(pq) + S_z^*(pq)S_x(pq)] \\
& + (1 - \sin^2\theta\sin^2\varphi)g_y^2 S_y^*(pq)S_y(pq) \\
& + \cos\theta\sin\theta\sin\varphi\, g_y g_z [S_y^*(pq)S_z(pq) + S_z^*(pq)S_y(pq)] \\
& + \sin^2\theta g_z^2 S_z^*(pq)S_z(pq) \}
\end{aligned}
\tag{8.25}
$$

In terms of the direction cosines l_i the intensity expressions in Equations 8.19 and 8.25 are

$$I_\parallel(p,q) = \sum_{m,n} l_m g_m S_m^*(pq) l_n g_n S_n(pq) \tag{8.26}$$

and

$$I_\perp(p,q) = (1/2)\left\{ \sum_n \left(1 - l_n^2\right) g_n^2 S_n^*(pq) S_n(pq) - \sum_{n \neq m} l_m l_n g_m g_n S_m^*(pq) S_n(pq) \right\} \tag{8.27}$$

Equations 8.26 and 8.27 were originally derived in a study on $S = 1$ organic biradi-cals by Wasserman et al., but in this early work the g-values were omitted and the prefactor 1/2 in Equation 8.27 was missing due to an integration error (Wasserman et al. 1964); these shortcomings have propagated in more recent literature, for example, Morrin and Bonnin 1999. Later attempts to derive general intensity expression contain typographical or mathematic errors, e.g., $g_y\cos\theta\sin\theta S_y$ instead of $g_y\cos\theta\sin\varphi S_y$ (van Veen 1978: Equation A6), or $k_y = \cos\alpha\sin\theta\cos\varphi - \cos\alpha\sin\varphi$ and $k_z = -\sin\theta\cos\varphi$, instead of the proper expressions in Equation 8.20 (Stevenson et al. 1986).

Let us now illustrate the use of intensity Equations 8.26 and 8.27 on the example of $S = 3/2$ for which we developed the energy matrix in Equation 7.38 on the basis of the spin wavefunctions in Equation 7.35 via the spin operations in Equation 7.36. We reorder the initial wavefunctions as

$$\varphi_i = \{|+3/2\rangle; |-3/2\rangle; |+1/2\rangle; |-1/2\rangle\} \tag{8.28}$$

which does, of course, not change any of the spin operations in Equation 7.36, but it causes a reordering of the energy matrix to:

$$
\begin{array}{c|cccc}
\langle +3/2| & D+3G_z & 0 & \sqrt{3}G_- & \sqrt{3}E \\
\langle -3/2| & 0 & D-3G_z & \sqrt{3}E & \sqrt{3}G_+ \\
\langle +1/2| & \sqrt{3}G_+ & \sqrt{3}E & -D+G_z & 2G_- \\
\langle -1/2| & \sqrt{3}E & \sqrt{3}G_- & 2G_+ & -D-G_z
\end{array}
\tag{8.29}
$$

Upon diagonalization, the basis set of spin wavefunctions in Equation 8.28 change into

$$|q_1\rangle = c_{11}|+3/2\rangle + c_{21}|-3/2\rangle + c_{31}|+1/2\rangle + c_{41}|-1/2\rangle$$

$$|q_2\rangle = c_{12}|+3/2\rangle + c_{22}|-3/2\rangle + c_{32}|+1/2\rangle + c_{42}|-1/2\rangle$$

$$|q_3\rangle = c_{13}|+3/2\rangle + c_{23}|-3/2\rangle + c_{33}|+1/2\rangle + c_{43}|-1/2\rangle$$

$$|q_4\rangle = c_{14}|+3/2\rangle + c_{24}|-3/2\rangle + c_{34}|+1/2\rangle + c_{44}|-1/2\rangle \tag{8.30}$$

That is, in general, a state ket has the form

$$|q_q\rangle = c_{1q}|+3/2\rangle + c_{2q}|-3/2\rangle + c_{3q}|+1/2\rangle + c_{4q}|-1/2\rangle \tag{8.31}$$

It is important to realize that the coefficients c_{nq} may well be complex numbers and, since we want to calculate (cf. Equation 8.18) $\langle p|S_i|q\rangle$ terms, that the general form of a state bra is

$$\langle p_p| = c_{1p}^*\langle+3/2| + c_{2p}^*\langle-3/2| + c_{3p}^*\langle+1/2| + c_{4p}^*\langle-1/2| \tag{8.32}$$

As always, to obtain the results of S_x and S_y operators we first apply the ladder operators S_+ and S_-, for example,

$$\langle p_p|S_-|q_q\rangle = \langle p_p|S_- c_{1q}|+3/2\rangle + \langle p_p|S_- c_{2q}|-3/2\rangle$$
$$+ \langle p_p|S_- c_{3q}|+1/2\rangle + \langle p_p|S_- c_{4q}|-1/2\rangle \tag{8.33}$$

and after working out the individual spin operations (which we already did in Equation 7.36d) the results can be combined with the proper part of $\langle p_p|$. Thus, the individual terms in Equation 8.33 become

$$\langle p_p|S_- c_{1q}|+3/2\rangle = \langle p_p|\sqrt{3}c_{1q}|+1/2\rangle = c_{3p}^*\langle+1/2|\sqrt{3}c_{1q}|+1/2\rangle = \sqrt{3}c_{3p}^* c_{1q}$$
$$\langle p_p|S_- c_{2q}|-3/2\rangle = \langle p_p|0 = 0$$
$$\langle p_p|S_- c_{3q}|+1/2\rangle = \langle p_p|2c_{3q}|-1/2\rangle = c_{4p}^*\langle-1/2|2c_{3q}|+1/2\rangle = 2c_{4p}^* c_{3q} \tag{8.34}$$
$$\langle p_p|S_- c_{4q}|-1/2\rangle = \langle p_p|\sqrt{3}c_{4q}|-3/2\rangle = c_{2p}^*\langle-3/2|\sqrt{3}c_{4q}|-3/2\rangle = \sqrt{3}c_{2p}^* c_{4q}$$

We thus obtain

$$\langle p_p|S_-|q_q\rangle = \sqrt{3}c_{3p}^* c_{1q} + 2c_{4p}^* c_{3q} + \sqrt{3}c_{2p}^* c_{4q}$$
$$\langle p_p|S_+|q_q\rangle = \sqrt{3}c_{4p}^* c_{2q} + \sqrt{3}c_{1p}^* c_{3q} + 2c_{3p}^* c_{4q} \tag{8.35}$$
$$\langle p_p|S_z|q_q\rangle = (3/2)c_{1p}^* c_{1q} - (3/2)c_{2p}^* c_{2q} + (1/2)c_{3p}^* c_{3q} - (1/2)c_{4p}^* c_{4q}$$

and with the previously (Equation 7.15) defined $S_x = (S_+ + S_-)/2$ and $S_y = (S_+ - S_-)/(2i)$, and using (cf. Equations 7.3 and 7.4) $(a + ib)^* = (a - ib)$ to obtain all the complex conjugate $\langle p_p|S_i^*|q_q\rangle$ terms from the $\langle p_p|S_i|q_q\rangle$ terms by proper sign changes, we have now all the terms worked out to be put into the intensity expressions 8.26 and 8.27.

8.4 NUMERICAL IMPLEMENTATION OF DIAGONALIZATION SOLUTIONS

Numerical procedures for the diagonalization and eigenvector calculation of matrices all trace back to EISPACK, a freely available software collection of subroutines originally written in double-precision FORTRAN77, but now also available as

FORTRAN90 updates, and with many descendants in FORTRAN and other higher languages. The three original EISPACK subroutines required for EPR energy matrices are htridi.f (reduction of Hermitian matrix), tql2.f (all eigenvalues and eigenvectors, symmetric tridiagonal matrix), and htribk.f (back-transformation of Hermitian matrix).

Diagonalization and extraction of eigenvector coefficients are CPU-intensive operations and problems of more than trivial complexity can easily turn out to be practically unsolvable. This is where EPR spectroscopy becomes a "sporting" method, and where bioEPR can enormously benefit from biochemical intuition. The central question to address is to what extent a complex biomolecular system can be strait-jacketed through artificially enforced symmetry without compromising its information content. If we are prepared to give up rigorous links between spin Hamiltonian parameters and the detailed electronic structure of the biomolecule, then what *do* we still want to be able to extract from a less-than-rigorous spectral analysis? I would venture the all-overriding criterion to be biochemical relevance. For example, if the X-band EPR spectrum of a copper protein is to be used as a fingerprint for the determination of its approximate ligand environment by means of a phenomenological truth table of A_z versus g_z values, then the question to what extent these values are only apparent (nondiagonal) due to tensor noncolinearity, represents an unnecessarily high level of sophistication. An approximate simulation based on analytical expressions for orthorhombic systems (cf. Equation 5.18), or even simply estimating apparent A and g-values from a read-out of the spectrum, will suffice. And if the required information concerns a spin count of how much of the copper is actually in the cupric form, then A-values are irrelevant, and an estimate of the g-values can be very approximate without affecting the overall uncertainty in the spin concentration determination. On the other hand, biochemically meaningful information on multiplicity and stoichiometry of paramagnetic prosthetic groups in complex enzymes requires a physically meaningful deconvolution of sumspectra and, therefore, a reasonably detailed understanding of the shape of the individual EPR powder patterns, which may require significant insight in the effects of (lack of) symmetry on these patterns.

As another example, when for a high-spin system subject to the intermediate-field condition, with a complex EPR spectrum, the valence state is to be unambiguously determined, a full-blown matrix diagonalization and state-vector determination may well be unavoidable (following data collection at multiple microwave frequencies and modes). Alternatively, a shrewd wet-lab experimentalist may altogether avoid this tedious and time-consuming analysis by trying to simply oxidize or reduce the preparation under study to an $S = 1/2$ state with a frequently much more easily and straightforwardly analyzable spectrum. However, if such an escape route is not a chemical option, then the following section applies.

In order to calculate the intensity of a specific transition by means of Equation 8.26 or Equation 8.27 we must choose values for the two subscript indices of the initial state $\langle p_p|$ and the final state $|q_q\rangle$. In our example of $S = 3/2$ the indices run over 1–4 (the spin multiplet has four sublevels) and so six different $p \leftrightarrow q$ transitions are possible: $1 \leftrightarrow 2$, $1 \leftrightarrow 3$, $1 \leftrightarrow 4$, $2 \leftrightarrow 3$, $2 \leftrightarrow 4$, and $3 \leftrightarrow 4$. We rearranged the $S = 3/2$ energy matrix of Equation 7.38 into the form given in Equation 8.29, and for the latter we worked out the terms $\langle p_p|S_i|q_q\rangle$ for the intensity expressions. This switching of matrix form holds a warning: the coefficients of the worked-out $\langle p_p|S_i|q_q\rangle$ terms adhere to a numbering scheme that applies to the matrix in Equation 8.29, but not to the matrix in Equation 7.38. Succinctly keeping track of the coefficient indices is mandatory. This also holds for the relative energy values of the substates: diagonalization of the matrix in Equation 7.38 versus the one in Equation 8.29 will afford the energy of the ground state at a different position of the diagonal of each matrix. Fortunately, in the diagonalization subroutines of the EISPACK suite and its descendants this bookkeeping is normally taken care of by means of energy sorting: the energy eigenvalues are provided in an indexed array in order of increasing value, that is, the ground-state energy is the first (or the last) element of the array. The corresponding eigenfunctions $|q_q\rangle$ are likewise ordered.

In Chapter 5 analytical expressions for the resonance condition were given in closed form, for example, (Equation 5.4) $B_{res} = g\beta/h\upsilon$. In the energy matrix diagonalization method we only obtain energy differences ΔE, and we must then go on and search the spin Hamiltonian parameter space, for example, $\{g, D, A, l_i, B\}$ for molecular orientations that afford a ΔE value such that $h\upsilon \approx \Delta E$. In practice, we define a field sweep, for example 0–5000 gauss, which we digitize in, for example, $n = 1024$ B_n-values (i.e., with 1023 equidistant intervals of 5000/1023 gauss), and then we diagonalize the matrix for *each* of these B_n-values and search for minimum values of the quantity $|\Delta E(B_n)-h\upsilon|$, which can be further minimized by interpolation between two discrete B_n-values. We can try a shortcut by starting with a rough grid of, say, 16 or 32 (instead of 1024) equidistant B_n-values to find out in what grid interval $\Delta E(B_n)-h\upsilon$ changes sign (i.e., in what grid interval the transition occurs). However, the course of energy levels with increasing magnetic field can be quite complicated. In particular, it may well happen that over the explored field range two energy levels wobble around in such a way that their mutual distance becomes equal to $h\upsilon$ on *more than one* occasion, that is, that *several* transitions (EPR spectra) occur between a single pair of levels.

For matrices of modest dimensions 1024 matrix diagonalizations may not be a serious CPU problem for a PC, but if we include (as we will in the next chapter) distributions in the spin Hamiltonian parameters the required CPU time goes up by, say, two orders of magnitude, and if we want to implement automatic minimization, we must pay with another two or three orders of magnitude in CPU-time.

8.5 A BRIEF ON PERTURBATION THEORY

In Chapter 5 we had a look at resonance conditions for a number of common spin systems in the form of analytical expressions. The use of these expressions in (numerical) analysis of spectra is much (up to many orders of magnitude) faster than the matrix algebra employed in the present chapter. However, these expressions have boundary conditions that limit their applicability: they are only valid in either strong-field or weak-field limits, that is, when the electronic Zeeman interaction is either dominating all other interactions ($B*S \gg X*Y$) or when the Zeeman term is itself a small perturbation to the zero-field interaction ($B*S \ll S*S$). Under intermediate-field conditions ($B*S \approx X*Y$) energy matrix diagonalization is called for. There is, however, a "gray area" where analytical expressions may or may not apply depending on how "good" the expressions are.

Analytical resonance conditions are expansions in terms of corrections to a "zeroth-order" expression. The more expansion terms are taken into account, the more precise the analytical expression approaches the exact solution, but at the same time: the more unwieldy the expression becomes. In practice, only first-order, second-order, and occasionally third-order corrections are included. The terms are derived with a tool from quantum mechanics known as *perturbation theory*. The general procedure is as follows: The interactions in the spin Hamiltonian are divided into a dominant one, H^0, and one or more weak perturbations, H'. The energy matrix for the dominant interaction is written out and diagonalized providing the zeroth-order energies E_i^0. The remaining term(s), the perturbing Hamiltonian, H', is applied to the new basis set, ψ_i, and the resulting elements $H'_{ij} = \langle \psi_i | H' | \psi_j \rangle$ can be placed in the energy matrix. Perturbation theory then provides the energies corrected up to second order as

$$E_i \simeq E_i^{(0)} + E_i^{(1)} + E_i^{(2)} \tag{8.36}$$

in which

$$E_i^{(1)} = H'_{ii}$$
$$E_i^{(2)} = -\sum_{j \neq i} \frac{H'_{ji} H'_{ij}}{E_j^{(0)} - E_i^{(0)}} \tag{8.37}$$

In other words, the diagonal elements of the perturbing Hamiltonian provide the first-order correction to the energies of the spin manifold, and the nondiagonal elements give the second-order corrections. Perturbation theory also provides expressions for the calculation of the coefficients of the second-order corrected wavefunctions $|\psi_i\rangle$ in terms of the original wavefunctions $|\varphi_i\rangle$

$$|\psi_i\rangle = |\varphi_i\rangle - \sum_{j \neq i} \frac{H'_{ji}}{E_j^{(0)} - E_i^{(0)}} |\varphi_j\rangle \tag{8.38}$$

which may be used for transition-probability calculations. Frequently, however, the zeroth-order intensity expression is taken to be sufficiently accurate, unless the perturbations have made originally forbidden transitions weakly allowed.

An—at least, theoretically—simple example is the $S = 1$ system in weak-field subject to a dominant zero-field interaction and a weakly perturbing electronic Zeeman interaction (similar to the $S = 2$ case treated above). The initial basis set is

$$\xi_1 \equiv |1^{sym}\rangle = \frac{1}{\sqrt{2}}(|+1\rangle + |-1\rangle)$$
$$\xi_2 \equiv |1^{anti}\rangle = \frac{1}{\sqrt{2}}(|+1\rangle - |-1\rangle) \tag{8.39}$$
$$\xi_3 = |0\rangle$$

and the energy matrix is

$$
\begin{array}{r|ccc}
\langle 1^{sym}| & D/3+E & 2G_z & 2G_x \\
\langle 1^{anti}| & 2G_z & D/3-E & i2G_y \\
\langle 0| & 2G_x & -i2G_y & -2D/3
\end{array}
\tag{8.40}
$$

This matrix is diagonal in the zero-field interaction, so the zeroth-order energy levels can be directly seen to be

$$
E_i^{(0)} = D/3+E
$$

$$
E_j^{(0)} = D/3-E
\tag{8.41}
$$

$$
E_k^{(0)} = -2D/3
$$

The perturbing Zeeman interaction has no elements on the diagonal, so there is no first-order correction. The second-order corrections are

$$
E_i^{(2)} = -\frac{2G_z 2G_z}{(D/3-E)-(D/3+E)} - \frac{2G_x 2G_x}{-2D/3-(D/3+E)} = \frac{4G_z^2}{2E} + \frac{4G_x^2}{D+E}
$$

$$
E_j^{(2)} = -\frac{2G_z 2G_z}{(D/3+E)-(D/3-E)} - \frac{i2G_y(-i2G_y)}{-2D/3-(D/3-E)} = -\frac{4G_z^2}{2E} + \frac{4G_y^2}{D-E}
\tag{8.42}
$$

$$
E_k^{(2)} = -\frac{2G_x 2G_x}{(D/3+E)-(-2D/3)} - \frac{i2G_y(-i2G_y)}{(D/3-E)-(-2D/3)} = -\frac{4G_x^2}{(D+E)} - \frac{4G_y^2}{(D-E)}
$$

and for the non-Kramer's doublet the intradoublet resonance condition becomes

$$
h\nu = (E_i^{(0)} + E_i^{(2)}) - (E_j^{(0)} + E_j^{(2)})
$$

$$
= (D/3+E)+4G_z^2/2E+4G_x^2/(D+E)-(D/3-E)+4G_z^2/2E-4G_y^2/(D-E)
$$

$$
= 2E+4G_z^2/E+4G_x^2/(D+E)-4G_y^2/(D-E)
\tag{8.43}
$$

In the z-direction this gives

$$
h\nu = 2E+4G_z^2/E
\tag{8.44}
$$

leading to the resonance condition

$$
4G_z^2 = h\nu E - 2E^2
$$

$$
B_{res} = \sqrt{h\nu E - 2E^2}/g_z l_z \beta
\tag{8.45}
$$

This is a pretty unusual expression, and it should warn us that resonance conditions derived via perturbation theory should always be checked for their validity under actual conditions. Suppose that the zero-field intradoublet splitting, $E^{(0)}(|1^{sym}\rangle)-E^{(0)}$

$(|1^{anti}\rangle) = 2E$, would be coincidentally approximately equal to the microwave quantum (so in X-band $2E \approx 0.3$ cm^{-1}), then we would expect—in parallel-mode EPR—to observe a resonance line very close to zero field. In this rather unlikely situation the electronic Zeeman term is indeed very small because $B \approx 0$, and Equation 8.45 should be valid. However, at increased field values G_z and E will become comparable, and so the derivation cannot be justified any longer.

More generally, in cases of small, or no rhombicity, $E \approx 0$, the non-Kramer's doublet levels are degenerate in zero field, and, therefore, the denominator in Equation 8.44 becomes zero, and the perturbation treatment breaks down completely. Two-fold degeneracy is, of course, very common in spin manifolds, and the perturbation theory is required for two-fold degenerate states, that is, with an additional step, namely the diagonalization of the 2×2 energy submatrix between the two degenerate levels and an associated adjustment of spin wavefunctions, followed by application of the regular perturbation approach.

In the present case we have to solve (cf. Equation 7.24)

$$\begin{vmatrix} D/3 + E - \lambda & 2G_z \\ 2G_z & D/3 - E - \lambda \end{vmatrix} = 0 \qquad (8.46)$$

which gives

$$E^{(1)}_{sym;anti} = D/3 \pm \sqrt{E^2 + 4G_z^2} \qquad (8.47)$$

and the energy difference

$$\Delta E = E^{(1)}_{sym} - E^{(1)}_{anti} = h\nu = 2\sqrt{E^2 + 4G_z^2} \qquad (8.48)$$

hence, the resonance condition

$$B_{res} = \sqrt{(h\nu)^2 - 4E^2} / (2g_z l_z \beta) \qquad (8.49)$$

which for $E \to 0$ gives an effective g-value $g_z^{eff} = 2g_z \approx 4$ as it should for $S = 1$ (cf. Figure 5.16). And for increasing E-value, the line moves towards zero field.

Diagonalization of the submatrix in Equation 8.46 affords the new subbasis set (cf. Equation 7.30)

$$|+\rangle = \cos\omega |1^{sym}\rangle + \sin\omega |1^{anti}\rangle$$
$$|-\rangle = -\sin\omega |1^{sim}\rangle + \cos\omega |1^{anti}\rangle \qquad (8.50)$$

defined by (cf. Equation 7.31)

$$\tan 2\omega = -2G_z / E$$
$$2\cos^2 \omega = 1 + [1 + (2G_z / E)^2]^{-0.5} \qquad (8.51)$$
$$2\sin^2 \omega = 1 - [1 + (2G_z / E)^2]^{-0.5}$$

And this redefines the complete energy matrix as

$$
\begin{array}{r|ccc}
\langle +| & D/3+\sqrt{E^2+4G_z^2} & 0 & \cos\omega 2G_x + i\sin\omega 2G_y \\
\langle -| & 0 & D/3-\sqrt{E^2+4G_z^2} & -\sin\omega 2G_x + i\cos\omega 2G_y \\
\langle 0| & \cos\omega 2G_x - i\sin\omega 2G_y & -\sin\omega 2G_x - i\cos\omega 2G_y & -2D/3
\end{array} \qquad (8.52)
$$

which can now be subjected to standard (i.e., nondegenerate) perturbation theory (as in Equations 8.41–8.45) to obtain all the energy levels of the spin multiplet correct to second order,

$$
E_i^{(2)} = -\frac{\cos^2\omega(2G_x)^2 + \sin^2\omega(2G_y)^2}{-(2D/3)-[(D/3)+\sqrt{E^2+(2G_z)^2}]}
$$

$$
E_j^{(2)} = -\frac{\sin^2\omega(2G_x)^2 + \cos^2\omega(2G_y)^2}{-(2D/3)-[(D/3)-\sqrt{E^2+(2G_z)^2}]} \qquad (8.53)
$$

$$
E_k^{(2)} = -\frac{\cos^2\omega(2G_x)^2 + \sin^2\omega(2G_y)^2}{[(D/3)+\sqrt{E^2+(2G_z)^2}]+(2D/3)} - \frac{\sin^2\omega(2G_x)^2 + \cos^2\omega(2G_y)^2}{[(D/3)-\sqrt{E^2+(2G_z)^2}]+(2D/3)}
$$

in which we did not work out the denominators, so that their origin from zero-order energy splittings remains clear.

From the first two $E^{(2)}$'s we obtain the resonance condition for the non-Kramer's doublet:

$$
h\nu = (D/3)+\sqrt{E^2+(2G_z)^2} - (D/3)+\sqrt{E^2+(2G_z)^2} + E_i^{(2)} - E_j^{(2)} \qquad (8.54)
$$

which gives in the z-direction (i.e., $l_z = 1$, $l_x = l_y = 0$) the simple expression

$$
B_{z-res} = \sqrt{(h\nu)^2 - 4E^2} / (2g_z\beta) \qquad (8.55)
$$

In the other two directions, very long expressions result unless we make the approximation that $D \gg \sqrt{E^2+(2G_z)^2}$, which makes the denominators of $E_i^{(2)}$ and $E_j^{(2)}$ identical, and we get

$$
B_{i-res} = \sqrt{(h\nu - 2E)/D} / (g_i\beta) \qquad i = x,y \qquad (8.56)
$$

A possible application of Equations 8.55 and 8.56 would be the analysis of EPR from high-spin Ni^{II} in model compounds and in nickel proteins (the latter is yet to be reported).

The last approximating step leading to Equation 8.56 illustrates the dilemma that one faces in the application of perturbation theory. The resulting equations are analytical, which in principle provides insight by inspection into the physical causes (e.g., rhombicity) of spectral positions, and which make for fast computer simulators, but these advantages may be largely lost in the complexity of the resulting resonance expressions. Traditionally, EPR texts abound with perturbation-theory-derived expressions, but ever-increasing computational capacity and availability makes the alternative approach of simply diagonalizing energy matrices and inspecting the resulting spectra an attractive alternative.

We have illustrated with this relatively simple example of an isolated $S = 1$ system how to apply nondegenerate and degenerate perturbation theory to develop approximate (up to second order) expressions for the resonance conditions of systems subject to interactions that can be divided into strong and weak terms. Further refinements to higher order can be made, but this is not frequently done where the expressions become rather complex. An example of possible biological relevance can be found in (Kirkpatrick et al. 1964; Von Waldkirch et al. 1972), namely high-spin Fe^{III} subject to strong zero-field interaction with Zeeman perturbation to third order. Also, the coefficients of new state vectors can be used to compute corrected transition probabilities, but these corrections are frequently considered insignificant (e.g., for the hyperfine expressions of $S = 1/2$ systems in Equations 5.18–5.22, or for the Zeeman expressions in Section 5.6) for effective $S = 1/2$ subspectra from half-integer high-spin systems.

9 Conformational Distributions

This chapter considers the distribution of spin Hamiltonian parameters and their relation to conformational distribution of biomolecular structure. Distribution of a g-value or "g-strain" leads to an inhomogeneous broadening of the resonance line. Just like the g-value, also the linewidth, W, in general, turns out to be anisotropic, and this has important consequences for "powder patterns," that is, for the shape of EPR spectra from randomly oriented molecules. A statistical theory of g-strain is developed, and it is subsequently found that a special case of this theory (the case of full correlation between strain parameters) turns out to properly describe broadening in bioEPR. The possible cause and nature of strain in paramagnetic proteins is discussed.

9.1 CLASSICAL MODELS OF ANISOTROPIC LINEWIDTH

In Chapter 4 we discussed inhomogeneous broadening as a reflection of conformational distribution. We noted that to derive the exact form of an inhomogeneous distribution would require detailed and complex information on the distribution of structure around the paramagnet, plus a way to translate this data into a distribution of g-values. As this information for biomolecules is generally not available—or has at least not been explored in any depth—we opted for the ad hoc description of a gaussian distribution with subsequent corroboration by comparison with experiment. However, with the introduction in Chapter 5 of anisotropy in g, or in any other spin Hamiltonian parameter, we have created a new problem: if the standard deviation of the gaussian is a reflection of the width of the distribution in g, and if g is anisotropic, then shouldn't the inhomogeneous EPR linewidth also be anisotropic? An overwhelming amount of experimental data clearly shows that this is indeed the case. And this then brings us to the problem addressed in this chapter: to understand the details of biomolecular EPR powder spectra requires a quantitative description of linewidth *anisotropy*, that is, the way the gaussian standard deviation depends on the orientation of the biomolecule in the external magnetic field.

As a starting point let us be faithful to the history of the subject and try a simple physical model due to (Johnston and Hecht, 1965); if the inhomogeneous EPR line reflects a distribution in g-values, then the anisotropy in the linewidth should be scalable to the anisotropy in the g-value. In other words, the analytical expression for g-anisotropy in terms of direction cosines, l_i, between B and the molecular xyz-axes system (Equation 5.5) leads to an equivalent expression for the linewidth W:

$$W = \sqrt{l_x^2 W_x^2 + l_y^2 W_y^2 + l_z^2 W_z^2} \tag{9.1}$$

In mathematical terms: we have defined a second-rank tensor W^2 that diagonalizes in the same axis system as the g^2 tensor. Note that we use W and not the gaussian standard deviation, σ, to keep generality in the shape of the distribution, that is, to allow for the possibility that the distribution may not turn out to be exactly gaussian after all. We define W as the half width at half height (HWHH) of the distribution (Figure 9.1). Be aware that in the literature several symbols and definitions are in use, and that frequently the definition is implicit or even ambiguous: symbols W, Γ, σ, Δ, etc., can be full width at half height (FWHH), or HWHH, or standard deviation, or any other intrinsic distribution parameter. Comparison of literature values or reproduction of published analyses must be preceded by a resolution of used definitions.

Figure 9.2A gives an example of an $S = 1/2$ spectrum simulated with the use of Equation 9.1. The simulation approximately fits the experimental $[2Fe-2S]^{1+}$ spectrum, but not exactly. In particular, the three spectral features around g_x, g_y, and g_z in the experimental spectrum exhibit asymmetries that are not reproduced in the simulation. From a biochemical point of view, the approximate fit is acceptable for certain purposes, but not acceptable for other goals: if double integration of the experimental spectrum (Chapter 6) for determining spin concentration would be difficult, because the spectrum would be noisy, or would be disturbed by an interfering radical signal, or by a poor baseline, for example, from a dirty resonator, then one could use the approximate simulation for the spin counting with an acceptable error (the difference in shape between the simulation and the experimental spectrum will not lead to an integration error greater than circa 10–15%). However, if the goal of spectral analysis

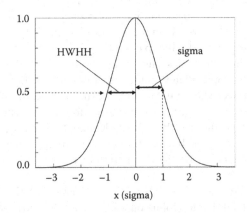

FIGURE 9.1 Linewidth versus standard deviation. A gaussian distribution of unit amplitude is plotted on an x-axis scale in units of the standard deviation (or: sigma). At $3.15 \times \sigma$ the unit intensity has dropped to 0.1%. The linewidth in simulations is usually expressed as the half-width at half height (HWHH), which is equal to circa $1.17 \times \sigma$, or as twice this value that is, the full width at half height (FWHH).

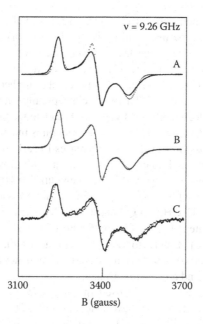

v = 9.26 GHz

A

B

C

3100   3400   3700

B (gauss)

FIGURE 9.2 EPR powder pattern of the [2Fe-2S]$^{1+}$ cluster in spinach ferredoxin. Trace A shows an attempt to fit the spectrum with the "diagonal" linewidth Equation 9.1. In trace B the spectrum is fitted with the "nondiagonal" g-strain Equation 9.18. Trace C shows an experiment in which the spectral features are slightly shifted (solid trace) under the influence of an external hydrostatic stress. (Data replotted from Hagen and Albracht 1982.)

would be to determine how many different prosthetic groups contribute to the overall spectrum, then the misfit would be unacceptable: the asymmetric "shoulders" in the experimental spectrum would have to be interpreted in terms of a second component (or even multiple components), while there are no indications of multiplicity from other experimental approaches—for example, crystallography, other spectroscopies, column chromatography, ultracentrifugation. Furthermore, the linewidths used in the simulation of Figure 9.2 do not really scale at all with the anisotropic deviation from the free electron g-value, so the ad hoc definition of Equation 9.1 does not appear to have any readily identifiable physical basis.

An alternative model (Venable 1967) proposes that the main cause of inhomogeneous broadening is unresolved superhyperfine interactions and, therefore, that the linewidth expression should be equivalent to the Equation 5.12 for the angular dependence of first-order hyperfine splitting:

$$W = \sqrt{l_x^2 g_x^4 W_x^2 + l_y^2 g_y^4 W_y^2 + l_z^2 g_z^4 W_z^2} \, / \, g^2 \tag{9.2}$$

For the spectrum in Figure 9.2 this proposal does not lead to an improved fit. In fact, a simulation based on Equation 9.2 is indistinguishable from that based on Equation 9.1. Perhaps the shape of the expression is too simple, and we should include g/A-tensor noncolinearity low-symmetry effects as in Section 5.5. The physical basis of this model is also disputable: superhyperfine splitting is usually not very anisotropic. Moreover, many metal ions in prosthetic groups lack first sphere coordination ligand atoms with a nuclear spin. In the example of Figure 9.2, the Fe ions are surrounded by S (99.2% ^{32}S: $I = 0$) and so the broadening should stem from second or higher sphere ^{14}N and ^{1}H and from rare isotopes like ^{13}C and ^{33}S. There is a conceptually simple, but experimentally involved test to check for significant hyperfine broadening: Since $S*B$ interaction is linear in the external magnetic field and $S*I$ interaction is independent of the magnetic field, W (in units of gauss) in Equation 9.1 increases linearly with increasing B, while W (again in units of gauss) in Equation 9.2 is invariant with increasing B. The result for such an experiment is given in Figure 9.3; the linewidth (FWHH) of the low-field peak of ferricytochrome c and of [2Fe-2S]$^{1+}$ ferredoxin are plotted as a function of microwave frequency from L-band (1 GHz) to Q-band (35 GHz). Clearly, the contribution from unresolved hyperfine interactions to the linewidth of ferredoxin has become near to negligible at X-band. Furthermore, to the—much broader—line from the cytochrome, there is no hyperfine contribution at all in the studied frequency range, even though five of the six first sphere ligand atoms are nitrogen.

Equations 9.1 and 9.2 are widely used in (bio)molecular EPR spectroscopy to describe linewidth anisotropy. We have just seen that both equations are theoretically unfounded and that in practice they fail to reproduce the details of experimental spectra. Apparently, neither unresolved ligand hyperfine splittings nor a distributed electron Zeeman interaction is the source of the observed inhomogeneous broadening. On the other hand, experiments like those in Figure 9.3 make it clear that the broadening, at least from X-band frequencies onwards, often is linear in the magnetic field. Although the cause of the broadening is not the Zeeman interaction, it manifests itself through its action on the Zeeman interaction. Based on this contention, and for lack of an established physical basis of the broadening, a statistical theory has been developed to formally describe its anisotropy, and thus to allow description (therefore: classification, deconvolution of components) of bioEPR powder patterns.

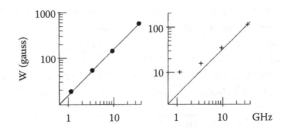

FIGURE 9.3 Linewidth as a function of microwave frequency. The linewidth (FWHH) of the low-field g_z-feature is plotted versus the frequency in L-, S-, X-, and Q-band. The left-hand panel is for the ferric low-spin heme in horse heart cytochrome c, and the right-hand panel is for the [2Fe-2S] cluster in spinach ferredoxin. (Data from Hagen 1989.)

9.2 STATISTICAL THEORY OF g-STRAIN

In the language of statistics, Equation 9.1 expresses the notion that the g-value is a random variable. The outcome of its measurement is slightly different for each individual molecule, and when we measure an ensemble of molecules, that is, a sample of finite size with an order-of-magnitude number of molecules approaching Avogadro's number, then we find a probability distribution of g-values over a domain of possible values (for a gaussian distribution, the domain formally runs from $-\infty$ to ∞) centered around an expected value, or mean, $\langle g \rangle$ (here, the g-value at the center of the gaussian). Equation 9.1 is a description of anisotropy in $W(g)$, and since g has three principal values, g_x, g_y, and g_z, the equation states the linewidth to reflect the statistical properties of three random variables with expected values $\langle g_x \rangle$, $\langle g_y \rangle$, and $\langle g_z \rangle$. Furthermore, since we found the scaling of W to the deviation of the g-value from the free electron value, $|g_i - g_e|$, not to be constant, the three random variables do not appear to be particularly strongly related to each other. This question to what extent random variables are related will prove to be extremely important in the EPR linewidth theory to be developed, so we'd better recall how "related" is defined in statistics.

Just as in everyday life, in statistics a relation is a pair-wise interaction. Suppose we have two random variables, g_a and g_b (e.g., one can think of an axial $S = 1/2$ system with $g_\parallel$ and $g_\perp$). The g-value is a random variable and a function of two other random variables: $g = f(g_a, g_b)$. Each random variable is distributed according to its own, say, gaussian distribution with a mean and a standard deviation, for g_a, for example, $\langle g_a \rangle$ and σ_a. The standard deviation is a measure of how much a random variable can deviate from its mean, either in a positive or negative direction. The standard deviation itself is a positive number as it is defined as the square root of the variance σ_a^2. The extent to which two random variables are "related," that is, how much their individual variation is intertwined, is then expressed in their covariance C_{ab}:

$$C_{ab} = \langle g_a g_b \rangle - \langle g_a \rangle \langle g_b \rangle \tag{9.3}$$

And a related way to express their interdependence is by means of their correlation coefficient r_{ab}:

$$r_{ab} = C_{ab} / (\sigma_a \sigma_b) \tag{9.4}$$

If two random variables are *un*correlated, then both their covariance C_{ab} and their correlation coefficient r_{ab} are equal to zero. If two random variables are *fully* correlated, then the absolute value of their covariance is $|C_{ab}| = \sigma_a \sigma_b$, and the absolute value of their correlation coefficient is unity: $|r_{ab}| = 1$. A key point to note for our EPR linewidth theory to be developed is that two fully correlated variables can be fully *positively* correlated: $r_{ab} = 1$, or fully *negatively* correlated: $r_{ab} = -1$. Of course, if two random variables are correlated "to some extent," then $0 < |C_{ab}| < \sigma_a \sigma_b$, and $0 < |r_{ab}| < 1$.

We know that a model in which the principal values of the g tensor are random variables, leading to Equation 9.1, falls short of describing experimental data in detail. Therefore, we now expand the model as follows: the random variables are the principal values of a physical entity that is characterized by a tensor in 3-D space, but whose nature and origin is yet to be identified. We assign the symbol p to this tensor and write the most general linear function of three random variables:

$$g = g^0 + R(\alpha,\beta,\gamma)pR^{-1}(\alpha,\beta,\gamma) \tag{9.5}$$

in which p is a tensor whose principal elements, p_1, p_2, p_3, are random variables, g^0 is a tensor whose elements do not fluctuate, and $R(\alpha,\beta,\gamma)$ is the 3-D rotation, which transforms the p principal axis system to the g^0 principal axis system. The random variables p_i have standard deviations σ_i, and their (possible) mutual correlations are described by pair-wise correlation coefficients, r_{12}, etc., and associated covariances $C_{12} = r_{12}\sigma_1\sigma_2$ etc. The expectation value $\langle g \rangle$ can then be developed in a Taylor series:

$$\langle g \rangle = g^0 + \frac{1}{2}\left(\sum_{i=1}^{n} \frac{\partial^2 g}{\partial p_i^2}\sigma_i^2 + 2\sum_{j>i}^{n} \frac{\partial^2 g}{\partial p_i \partial p_j} r_{ij}\sigma_i\sigma_j \right) \tag{9.6}$$

and its variance σ_g^2 is

$$\sigma_g^2 = \sum_{i=1}^{n}\left(\frac{\partial g}{\partial p_i}\right)^2 \sigma_i^2 + 2\sum_{j>i}^{3}\left(\frac{\partial g}{\partial p_i}\right)\left(\frac{\partial g}{\partial p_j}\right) r_{ij}\sigma_i\sigma_j \tag{9.7}$$

It is precisely this variance of $\langle g \rangle$ that we are after, because its square root gives us the angular dependent linewidth. A general expression in matrix notation can be derived for the variance (Hagen et al. 1985c):

$$\sigma_g^2 = {}_2\Lambda^{-1} \bullet {}_2D_2P_2D^{-1} \bullet {}_2\Lambda / g^2 \tag{9.8}$$

in which we have dropped the superscript "0" for the nonfluctuating g-value. The presubscript "2" means trouble: it indicates that the rotation in Equation 9.8 is not the "simple" 3-D rotation defined in Equation 8.13, but is in fact a rotation in an expanded space of dimensionality 5, known in group theory as "$D^{(2)}$" (ibidem). Thus, all the vectors and matrices in Equation 9.8 are 5-D. They are defined as:

$$_2\Lambda^{-1} = \left[\left(3g_z l_z^2 - 1\right)/2,\ \sqrt{3}\left(g_x l_x^2 - g_y l_y^2\right)/2,\ \sqrt{3g_x g_y}l_x l_y,\ \sqrt{3g_x g_z}l_x l_z,\ \sqrt{3g_y g_z}l_y l_z \right] \tag{9.9}$$

$$_2D^{(\alpha,0,0)} = \begin{pmatrix} 1 & 0 & 0 & 0 & 0 \\ 0 & 1-2\sin^2\alpha & -2\sin\alpha\cos\alpha & 0 & 0 \\ 0 & 2\sin\alpha\cos\alpha & 1-2\sin^2\alpha & 0 & 0 \\ 0 & 0 & 0 & \cos\alpha & -\sin\alpha \\ 0 & 0 & 0 & \sin\alpha & \cos\alpha \end{pmatrix} \tag{9.10}$$

with an equivalent expression for $_2D(0,0,\gamma)$, and

$$_2D(0,\beta,0) = \begin{pmatrix} 1-(3/2)\sin^2\beta & (\sqrt{3}/2)\sin^2\beta & 0 & -\sqrt{3}\cos\beta\sin\beta & 0 \\ (\sqrt{3}/2)\sin^2\beta & 1-(1/2)\sin^2\beta & 0 & \cos\beta\sin\beta & 0 \\ 0 & 0 & \cos\beta & 0 & \sin\beta \\ \sqrt{3}\cos\beta\sin\beta & -\cos\beta\sin\beta & 0 & 1-2\sin^2\beta & 0 \\ 0 & 0 & -\sin\beta & 0 & \cos\beta \end{pmatrix}$$

$$(9.11)$$

and

$$_2D(\alpha,\beta,\gamma) = {}_2D(\gamma,0,0)\,{}_2D(0,\beta,0)\,{}_2D(0,0,\alpha) \tag{9.12}$$

and, finally,

$$_2P =$$

$$\begin{pmatrix} \sigma_1^2 & \dfrac{\sigma_2^2-\sigma_1^2}{\sqrt{3}} & 0 & 0 & 0 \\ \dfrac{\sigma_2^2-\sigma_1^2}{\sqrt{3}} & \dfrac{2\sigma_1^2+2\sigma_2^2-\sigma_3^2}{3} & 0 & 0 & 0 \\ 0 & 0 & \dfrac{\sigma_1^2+r_{12}+\sigma_2^2-\sigma_3^2}{3} & 0 & 0 \\ 0 & 0 & 0 & \dfrac{\sigma_1^2+r_{13}+\sigma_3^2-\sigma_2^2}{3} & 0 \\ 0 & 0 & 0 & 0 & \dfrac{\sigma_2^2+r_{23}+\sigma_3^2-\sigma_1^2}{3} \end{pmatrix}$$

$$(9.13)$$

Now this Equation 9.8 is certainly not an expression that one would want to work out analytically, but coding it in a higher computer language should be straightforward. The EPR linewidth would then be defined by Equation 9.8 as the standard deviation σ_g, which is a function of the direction cosines, l_i, and of the fitting parameters σ_1, σ_2, σ_3, r_{12}, r_{13}, r_{23} (which together describe the joint distribution of the random variables p_1, p_2, and p_3), and α, β, γ. Application reveals a practical problem: variation in the angle-parameters, α, β, γ, leads to rapid changes in the values of σ_g along the g-tensor axes. Since the latter are what we observe as linewidth values of the main features in an experimental $S = 1/2$ powder spectrum, variation of fitting parameters is not easily monitored by visual inspection. The only option left is to let the computer do a systematic scan through the space of nine linewidth fitting parameters.

9.3 SPECIAL CASE OF FULL CORRELATION

Remarkably, application of the above theory (Hagen et al. 1985d; Hearshen 1986) has revealed that many spectra of biological $S = 1/2$ systems can be excellently fit by a special case of Equation 9.8, namely the case of full (positive or negative) correlation: $|r_{ij}| \equiv 1$. This unexpected finding leads to a linewidth equation that is not only much simpler than Equation 9.8, but also very much more easy to use in simulations.

We will now define this equation and discuss practical aspects of its implementation in computer simulators. Furthermore, in the next section we will develop a biomolecular interpretation of the fully correlated distributions of the p-tensor elements. For $|r_{ij}| \equiv 1$ Equation 9.8 simplifies to (Hagen et al. 1985c)

$$\sigma_g^2 = ({}_1\Lambda^{-1} \bullet {}_1D_1P_1D^{-1} \bullet {}_1\Lambda)^2 / g^2 \qquad (9.14)$$

with the row vector

$$_1\Lambda^{-1} = \left(l_x\sqrt{g_x}, \ l_y\sqrt{g_y}, \ l_z\sqrt{g_z} \right) \qquad (9.15)$$

and in which $_1D$ is the 3D rotation previously defined as R in Equation 8.13, and

$$_1P = \begin{pmatrix} r_{23}\sigma_1 & 0 & 0 \\ 0 & r_{13}\sigma_2 & 0 \\ 0 & 0 & r_{12}\sigma_3 \end{pmatrix} \qquad (9.16)$$

The three-matrix product $_1D_1P_1D^{-1}$ is a symmetrical matrix with real elements and can thus be written as

$$_1D_1P_1D^{-1} = \begin{pmatrix} \Delta_{xx} & \Delta_{xy} & \Delta_{xz} \\ \Delta_{xy} & \Delta_{yy} & \Delta_{yz} \\ \Delta_{xz} & \Delta_{yz} & \Delta_{zz} \end{pmatrix} \qquad (9.17)$$

but note that since in the $_1P$-matrix in Equation 9.16 $r_{ij} = \pm 1$, each Δ can be positive or negative. With this substitution the worked out Equation 9.14 becomes

$$\sigma_g = \left| l_x^2 g_x \Delta_{xx} + l_y^2 g_y \Delta_{yy} + l_z^2 g_z \Delta_{zz} \right.$$
$$\left. + 2l_x l_y \sqrt{g_x g_y} \Delta_{xy} + 2l_x l_z \sqrt{g_x g_z} \Delta_{xz} + 2l_y l_z \sqrt{g_y g_z} \Delta_{yz} \right| / g \qquad (9.18)$$

For the special case of a positive definite diagonal Δ-matrix (i.e., $\Delta_{ii} > 0$ and $\Delta_{ij} = 0$) Equation 9.18 has similarity to our starting-point Equation 9.1 for g as a random variable and also to Equation 9.2 for unresolved superhyperfine broadening. Indeed, a simulation based on this special form of Equation 9.18 is not discernable from one based on either Equation 9.1 or Equation 9.2. This indicates an important asset of Equation 9.18, namely, the fact that the diagonal Δ_{ii} values can be approximately identified with the experimentally observed linewidths along the g-tensor axes, which makes tracking of fitting routines practical. However, the full-blown Equation 9.18 differs from the early linewidth expressions Equations 9.1 and 9.2 in two essential aspects: (i) the Δ's can be positive or negative; and (ii) the linewidth Δ-matrix has off-diagonal elements. These two aspects cause the domain of possible spectral shapes that can be generated with Equation 9.18 to be much wider than that covered by the early equations. In particular, simulators that incorporate Equation 9.18 to describe linewidth anisotropy can generate the asymmetries that were revealed in, for example, the misfit in Figure 9.2A. An illustration of this capacity is given in Figure 9.2B. In geometrical terms this capacity to generate asymmetries can be traced back to the noncolinearity of the g- and p-tensors, which results in an angular dependence of the linewidth that can vary rapidly near the canonical orientations of the g-tensor.

Implementation of Equation 9.18 in spectral simulators requires some extra precautions (Hagen 1981; Hagen et al. 1985d): (A) The increased periodicity now requires one half of the unit sphere to be scanned. (B) The fact that the term within the absolute-value bars in Equation 9.18 can change sign as a function of molecular orientation implies the possibility that for specific orientations the linewidth becomes equal to zero. To avoid a program crash due to a zero divide, e.g., in the expression for the lineshape in Equation 4.8, a "residual" linewidth W_0 has to be introduced:

$$W = \sqrt{W_0^2 + W_g^2} \tag{9.19}$$

where the quadrature indicates the origin of W_0 to be different (independent) from that of W_g. One can think of many causes for W_0, such as unresolved superhyperfine splittings or even lifetime broadening, but in practice W_0 simply acts as a dummy to prevent division by zero. To this goal W_0 may be set equal to the digital resolution of the simulation. (C) Using a small value for W_0 increases the risk of mosaic artifacts (cf. Chapter 4) and, therefore, the use of Equation 9.18 calls for an increased number (typically by one or two orders of magnitude) of molecular orientations to be calculated, that is, for increased CPU time. (D) The effect of the distributed p-tensor is through the Zeeman interaction, so distributions are conveniently computed on a scale linear in the g-value. Therefore, simulators using Equation 9.18 (or Equation 9.8) generate distributions to compose powder spectra "in frequency space" or "in g-space" that is, on a linear g-scale, and only after the full powder EPR absorption spectrum has been obtained, it is transformed by interpolating bijection to "B-space" that is, to a linear field scale, whereupon it is finally differentiated to obtain the first-derivative field-scanned spectrum. The pseudo-code to get from an absorption spectrum of n points in g-space to a derivative spectrum of n points in B-space is:

```
INPUT: n-point array: Iₘ(n) (amplitudes of absorption spectrum in g-space)
INPUT: g-space limits: gₘᵢₙ, gₘₐₓ, gₛₜₑₚ = (gₘₐₓ - gₘᵢₙ) / (n-1)
INPUT: B-space limits: Bₘᵢₙ, Bₘₐₓ, Bₛₜₑₚ = (Bₘₐₓ - Bₘᵢₙ) / (n-1)
INPUT: microwave frequency ν
DO STEP in B from Bₘᵢₙ to Bₘₐₓ by Bₛₜₑₚ
    Compute g(B) (Equation 2.6)
    Retrieve the two Iₘ values at the discrete g-values flanking g(B)
    Interpolate and store as Iᵦ
    Multiply Iᵦ by g(B) for area normalization
END STEP in B
Compute derivative dIᵦ/dB
```

A simulator based on this algorithm and generating multiple $S = 1/2$ components in adjustable stoichiometries each inhomogeneously broadened according to Equation 9.18 is part of the program suite.

9.4   A (BIO)MOLECULAR INTERPRETATION OF g-STRAIN

Our finding that linewidth anisotropy in biomolecular EPR spectra can be described by a statistical theory in which the random variables that cause the broadening are fully correlated, does not only make analysis by simulation practical; it also holds a message on the nature of the ultimate source of the broadening: if the three principal elements of the p-tensor are fully correlated, then they should find their cause in a single, scalar quantity.

Fritz et al. were the first to propose the name "g-strain" for the inhomogeneous broadening of EPR from a ferredoxin in which "the iron site of each protein has a slightly different conformation which results in a distribution of g values and hence an apparent broadening of the EPR line" (Fritz et al. 1971). The word "strain" implies the existence of a stress exerted on a local body, that is, by the periphery of a metalloprotein on its metal coordination site, and in its turn of the surrounding medium on the external of the protein. Here, the medium is an aqueous solution or the vitreous or polycrystalline state afforded upon freezing. Support for this mechanical picture comes from an experiment (Figure 9.2C) that makes use of the volume expansion of water upon freezing, in which a plastic cell with a closing screw is maximally filled with a solution of ferredoxin (i.e., with minimal head space) and subsequently frozen in liquid nitrogen to create a superstressed polycrystalline sample: the features in the resulting EPR spectrum are slightly shifted from its reference spectrum taken with a sample in an open tube frozen at ambient pressure (Hagen and Albracht 1982). The extent of the shifts (typically circa 5 gauss) is comparable to those observed in the EPR of transition ions doped in inorganic single crystals, for example, MgO, subjected to uniaxial mechanical stresses of circa 500 kg/cm^2 (Feher 1964; Tucker 1966).

A stress that is describable by a single scalar can be identified with a hydrostatic pressure, and this can perhaps be envisioned as the isotropic effect of the (frozen) medium on the globular-like contour of an "entrapped" protein. Of course, transduction of the strain at the protein surface via the complex network of chemical bonds of the protein 3-D structure will result in a local strain at the metal site that is not isotropic at all. In terms of the spin Hamiltonian the local strain is just another "field" (or: operator) to be added to our small collection of "main players," B, S, and I (section 5.1). We assign it the symbol T, and we note that in three-dimensional space, contrast to B, S, and I, which are each three-component vectors. T is a symmetrical tensor with six independent elements:

$$T = \begin{pmatrix} T_{xx} & T_{xy} & T_{xz} \\ T_{xy} & T_{yy} & T_{yz} \\ T_{xz} & T_{yz} & T_{zz} \end{pmatrix} \tag{9.20}$$

namely, the normal-strain components, T_{ii}, and the shear-strain components, T_{ij}.

We can now extend the spin Hamiltonians by making combinations of T, with B, and/or S, and/or I, and since we are interested in the effect of strain on the g-value from the electronic Zeeman interaction ($B*S$), the combination of interest here is $T*B*S$.

In the previous two chapters we repeatedly discussed the necessity to make choices of simplification to keep spectral analysis tractable. These choices frequently come down to assuming a higher symmetry of the paramagnetic site than strictly compatible with structural information or with chemical intuition. Ignoring low-symmetry terms may be justified when their spectral effects can be argued to be significantly smaller than those of equivalent high-symmetry terms. An example is the retaining of the cubic $a_n(S_+^n + S_-^n)$ term (with n = even) for integer-spin systems while ignoring all other higher power terms in S (cf. Section 8.1). Here, we are facing the necessity of making a similar decision: in the absence of any symmetry (as, e.g., can be expected for the deformed quasi-tetrahedral coordination of iron in [2Fe-2S] ferredoxins) a complete description of the coupling between a six-component strain T and the two three-component vectors of the Zeeman interaction $B*S$ would require the addition of a total of $6 \times 3 \times 3 = 54$ $T*S*B$ terms to the spin Hamiltonian. This would obviously define a very highly underdetermined problem of analysis, where we found experimentally the g-strain to be describable by maximally only six Δ_{ij} linewidth parameters (Equation 9.18). We choose to once more take the approach that proved to be practical in the case of higher-power S-terms for integer-spin systems: we start with the terms required for cubic symmetry only (and we eventually decide to limit the analysis to these terms). In structural terms, the limitation means, for example, that for an iron site in a [2Fe-2S] ferredoxin we assume the coordination to be a perfect tetrahedron of four S atoms, and for the iron site in cytochrome c we assume a perfectly octahedral coordination by six N atoms.

The spin Hamiltonian encompassing electronic Zeeman plus strain interaction for a cubic $S = 1/2$ system is (Pake and Estle 1973: Equation 7–21):

$$H_S = g\beta B \bullet S + P_1(T_{xx} + T_{yy} + T_{zz})B \bullet S$$

$$+(2P_2 / 3)\sum_{x,y,z} T_{ii}(2B_iS_i - B_jS_j - B_kS_k) + P_3\sum_{x,y,z} T_{ij}(B_iS_j + B_jS_i) \qquad (9.21)$$

and the resulting effective g-value for a frequency-swept spectrum is given in (ibidem: Equation 7–22) as:

$$g_{eff} = g + (P_1 / \beta)(T_{xx} + T_{yy} + T_{zz})$$

$$+ (2P_2 / 3\beta)[T_{xx}(3l_x^2 - 1) + T_{yy}(3l_y^2 - 1) + T_{zz}(3l_z^2 - 1)]$$

$$+ (2P_3 / \beta)(T_{yz}l_yl_z + T_{zx}l_zl_x + T_{xy}l_xl_y) \qquad (9.22)$$

Rewriting for a field-swept spectrum gives

$$g_{eff} = g + (P_1 / \beta)(T_{xx} + T_{yy} + T_{zz})$$

$$+ (2P_2 / 3\beta)[T_{xx}(3g_xl_x^2 - 1) + T_{yy}(3g_yl_y^2 - 1) + T_{zz}(3g_zl_z^2 - 1)]$$

$$+ (2P_3 / \beta)\left(T_{yz}\sqrt{g_yg_z}\,l_yl_z + T_{zx}\sqrt{g_zg_x}\,l_zl_x + T_{xy}\sqrt{g_xg_y}\,l_xl_y\right) \qquad (9.23)$$

Note that the Zeeman interaction for a cubic system results in an isotropic g-value, but the combination with strain *lowers* the symmetry at least to axial (at least one of the $T_{ii} \neq 0$), and generally to rhombic. In other words, application of a general strain to a cubic system produces a symmetry identical to the one underlying a Zeeman interaction with three different g-values. In yet other words, a simple $S = 1/2$ system subject to a rhombic electronic Zeeman interaction only, can formally be described as a cubic system deformed by strain.

Rearrangement of Equation 9.23 separates isotropic shift from anisotropic shift:

$$g_{eff} = g + (1/\beta)(3P_1 - 2P_2)(T_{xx} + T_{yy} + T_{zz})/3$$
$$+ (1/\beta)2P_2(T_{xx}g_x l_x^2 + T_{yy}g_y l_y^2 + T_{zz}g_z l_z^2)$$
$$+ (1/\beta)2P_3\left(T_{yz}\sqrt{g_y g_z}\, l_y l_z + T_{zx}\sqrt{g_z g_x}\, l_z l_x + T_{xy}\sqrt{g_x g_y}\, l_x l_y\right) \qquad (9.24)$$

Comparing the strain-induced anisotropic part of the effective g-value in Equation 9.24 with the linewidth Equation 9.18 for fully correlated g-strain reveals a remarkable correspondence: they are equivalent (apart from a sign ambiguity) within the substitutions

$$\Delta_{ii} \leftrightarrow (1/\beta)2P_2 T_{ii}$$
$$\Delta_{ij} \leftrightarrow (1/\beta)2P_3 T_{ij} \qquad (9.25)$$

This suggests that the g-strain in EPR spectra contains detailed geometric and energetic information on the 3-D position of ligand atoms for a given metalloprotein conformation with respect to a (virtual) cubic coordination, and therefore, of position and interconversion energy (by strain) of the 3-D position of ligand atoms in, say, two different protein conformations. This is an unexplored area of research.

9.5   A-STRAIN AND D-STRAIN: COUPLING TO OTHER INTERACTIONS

Just like the Zeeman interaction ($S*B$), the hyperfine interaction ($S*I$) is a bilinear term and its coupling to strain ($T*S*I$), which we will call "A-strain" (also, "K-strain"), should be formally similar to the g-strain ($T*S*B$) just discussed. In the early work of Tucker on the effective $S = 1/2$ system Co^{2+} in the cubic host MgO, a shift in central hyperfine splitting was found to be proportional to the strain-induced g-shift given by Equation 9.22 (Tucker 1966).

A qualitatively important phenomenon is that the combined effects of g-strain and A-strain always result in different shifts for each individual hyperfine line, simply because the splitting from average g-value is a function of the m_I value (cf. Equation 5.10). For a given sign of the hyperfine splitting, say $A > 0$, the lines with $m_i > 0$ will experience a total shift greater than the g-shift only, while those with $m_I < 0$ will typically be subject to a shift less than the g-shift. For a broadening due to a distributed shift the result is a linewidth (and relative intensity) that varies with m_I. The effect is illustrated in Figure 9.4 on the parallel four-line pattern of the hydrated copper ion.

The figure also shows the effect of an externally applied hydrostatic pressure (using the method of Figure 9.2C): the two outermost lines are seen to be shifted in opposite direction (Hagen 1982a).

A.S. Brill is possibly the first to have identified the source of this effect when he simulated the X-band spectrum of Cu^{II} in plastocyanin using a gaussian distribution of spin Hamiltonian parameters (Brill 1977). The fit is qualitative only, and today, over three decades later, the situation is unchanged: no theory plus associated software is available to obtain truly quantitative simulations of X-band (and lower frequencies) powder EPR from $S = 1/2$ biomolecules and model systems subject to central hyperfine interaction. For a given molecular orientation, the dependence of linewidth on nuclear quantum number has been proposed to be (Bogomolova et al. 1978, Froncisz and Hyde 1980, Hagen 1981, respectively)

$$W = \sqrt{A + Bm_I + Cm_I^2} \tag{9.26a}$$

$$W = \sqrt{A^2 + 2\varepsilon ACm_I + C^2m_I^2}; \quad 0.8 < \varepsilon < 1 \tag{9.26b}$$

$$W = A + Bm_I + Cm_I^2 \tag{9.26c}$$

Qualitatively, all proposals indicate a linear dependence on m_I (linewidth over a hyperfine pattern increases from low to high field or vice versa; cf. Figure 9.4) plus a quadratic dependence on m_I (outermost lines more broadened than inner lines). Multiple potential complications are associated with the lump parameters A, B, C, notably, their frequency dependence (Froncisz and Hyde 1980), partial correlation with g-strain (Hagen 1981), and low-symmetry effects (Hagen 1982a). The bottom line: quantitative description of these types of spectra has been for quite some time, and still is, awaiting maturation.

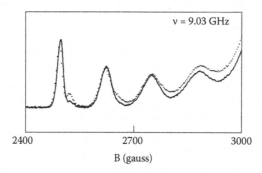

FIGURE 9.4 Effect of stress on a hyperfine pattern. The four-line parallel hyperfine pattern of the elongated CuO_6 octahedron in $^{63}Cu(H_2O)_6$ is shown in the presence (dotted line) and absence (solid line) of an external hydrostatic stress. (Modified from Hagen 1982a.)

Also, the zero-field interaction can be coupled to strain ($T*S*S$) giving rise to D-strain. Equations for high symmetry can be found in (Pake and Estle 1973: Chapter 7), but these are yet to be explored in bioEPR spectroscopy. More recent work has indicated that for half-integer high spin systems in the weak-field limit where the rhombograms apply, semiquantitative simulation of experimental data can be obtained assuming a simple Gaussian distribution in the rhombicity $\eta = E/D$. The practical value for such a simply one-parameter distribution model is in the fact that it "explains" (i.e., reproduces) otherwise unexplainable features in $S = n/2$ spectra such as the presence of strong $g = 4.3$ features even for systems of much less than maximal rhombicity and in the virtual absence of contaminating "dirty iron." The rhombicity-distribution model will be worked out and illustrated in Chapter 12 on high-spin systems, in Section 12.2.

Part 3

Specific Experiments

Having acquired a sound background in experimental and theoretical aspects of bioEPR spectroscopy, in Part 3 we take a closer look at a number of common experiments. We consider interactions between different centers; we also further advance into the realm of high-spin systems, then we develop protocols for EPR measurement of biochemically relevant thermodynamic and kinetic parameters, and we close off with some general thoughts on the planning of bioEPR experiments. We start off with the special subject of "room temperature" EPR: the study of samples in which the paramagnetic molecules are allowed a certain degree of rotational motion.

Biomolecular spectroscopy on frozen samples at cryogenic temperatures has the distinct disadvantage that the biomolecules are in a state that is not particularly "physiological." Recall that EPR spectroscopy is done at low temperatures to sharpen-up spectra by slowing down relaxation, to increase amplitude by increasing Boltzmann population differences, and to decrease diamagnetic absorption of microwaves by changing from water to ice. Certain $S = 1/2$ systems, notably radicals and a few mononuclear metal ions, have sufficiently slow relaxation, and sufficiently limited spectral anisotropy to allow their EPR detection in the liquid phase at ambient temperatures, be it in aqueous samples of reduced size.

10 Aqueous Solutions

Transition ions and radicals is what biomolecular EPR spectroscopy is all about. The abundance of high-spin configurations among transition ions is complemented by a relative paucity among radicals. Biologically relevant radicals are predominantly $S = 1/2$ systems; biradicals are rather less common (e.g., excited triplets in photosynthesis), and radicals with more than two unpaired electrons are not known to occur in living cells. Consequently, the EPR spectroscopy of radicals, both experimentally and theoretically, is typically relatively simple; however, the flip side of the coin frequently shows an experimentally complex chemistry. Key to this complexity are the high reactivity and/or the short lifetime of radicals, and cells amply employ these properties to their benefit by either assigning a biological function congruent with a rapid decay (radicals in signaling pathways) or by confining radical reactivity to a specific and well-controlled environment (radicals in enzyme active sites). Thus, the lifetime of a bioradical can be orders of magnitude less (the second messenger NO radical) or orders of magnitude greater (the flavin radical in the enzyme DNAlyase) than the typical throughput time of a cw-EPR experiment. The problem of a radical that reacts away before we have had a change to look at it with our spectrometer knows two conceptually and experimentally very different solutions: (1) chemical stabilization by letting the radical react with a diamagnetic compound to produce a different radical with a longer lifetime and (2) physical stabilization by freezing the radical and cooling it to a temperature below circa 175 K. The first approach is popular particularly in medically-oriented and also in food-related research, and it is discussed below. The latter approach is preferred by enzymologists, and it is discussed later in Chapter 13.

10.1 SPIN TRAPS

A spin trap is a diamagnetic compound that reacts with a radical by addition of the radical functionality typically to a double bond in the trap, thus forming a new radical that is more stable (better, less unstable) than the original radical. By far the most common class of spin traps are nitrone compounds that, upon addition of the primary radical, produce a stable aminoxyl radical (Figure 10.1). The compound DMPO is the paradigmatic spin trap; it is readily available, widely used, and its EPR spectra are relatively easy to interpret. Some of its radical adducts have impractically short lifetimes.

Spin traps are usually not practical to stabilize radicals on biomacromolecules because the reactivity is too low, presumably due to steric hindrance. Spin traps are used to stabilize physiologically relevant radicals of relatively small size such as hydroxyl, superoxide, and carbon-based radicals on organic molecules, for example, lipids.

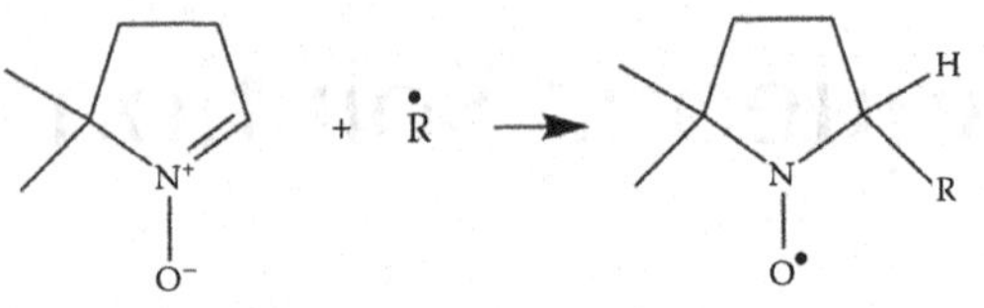

FIGURE 10.1 The structure of DMPO. The diamagnetic compound 5,5-dimethyl-1-pyrroline-N-oxide reacts with an unstable radical R• to form a relatively stable radical adduct.

From the EPR spectroscopist's viewpoint the spin-trap experiment is next to trivial; the molecular mass of the radical adduct is small enough to guarantee the molecule to tumble sufficiently rapidly at ambient temperatures in aqueous solution to ensure complete averaging away of any anisotropy in the spin Hamiltonian:

$$H_S = g\beta BS + \sum_i A_i SI_i \qquad (10.1)$$

The superhyperfine splittings are sufficiently small to ignore second-order effects at X-band, and for adducts of the nitrone compounds splitting from the nitrone-N and the beta-H are the only resolved hyperfine interactions, thus affording the extremely simple resonance condition (cf. Equation 5.10)

$$B_{res} = (1/g_{iso})(h\nu/\beta) - A_{iso}^N m_I^N - A_{iso}^H m_I^H \qquad (10.2)$$

in which the hyperfine splitting constants are in magnetic-field units (gauss). Using the capillary-in-a-regular-tube setup described in Chapter 3, retuning of the spectrometer is usually not required, thus allowing for a relatively fast throughput of samples. Also, as all spectra are rather similar, a single set of measuring conditions usually suffices, but note that microwave power levels above 20 milliwatt (−10 dB) should be avoided because they lead to significant warming-up of the sample through diamagnetic absorption by the water (the microwave-oven effect).

Equation 10.2 affords a single, isotropic line split into a triplet by the ^{14}N nucleus and a further splitting into doublets by the ^1H nucleus, altogether resulting in a pattern of six-lines of equal width and intensity. For particular magnitudes of A^N and A^H the lines may partly, or completely overlap as is the case for the HO• adduct of DMPO, which has $A^N \approx A^H$, resulting in a four-line pattern with 1:2:2:1 intensities (Figure 10.2).

The primary literature on spin trapping is diffused over many subfields of science, and its content also attests to the multiple and bewildering problems of chemical kinetics related to slow rate of addition, to radical-radical side reactions, to decay of primary radicals, to decay of adducts, to side reactions by light or by oxygen, to enzyme inhibition effects, to toxicity effects, and so on and so forth (e.g., Janzen 1995). In short, the EPR experiment is in principle extremely straightforward, but it may in practice be difficult due to low signal intensities and to multiple overlapping signals from decay products. However, the chemistry is the truly challenging part.

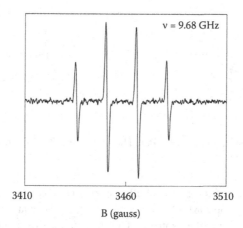

FIGURE 10.2 The spectrum of the DMPO •OH adduct. The rapidly tumbling adduct affords an isotropic spectrum split by ^{14}N ($I = 1$) in three lines, each of which is split by the β proton ($I = 1/2$) in two lines. Overlap of lines, due to $A_N \approx A_H$, gives a 1:2:2:1 intensity pattern.

Let us round off the subject by comparison of two extreme examples to demarcate the battleground of spin-trapping EPR practice. An easy experiment without pitfalls would be to assess the shelf life of a beer in a translucent bottle by means of the total EPR intensity of so-called "reactive oxygen species" radical adducts to a spin trap dissolved in the beer as a function of time. Shelf life here means how long it takes before the beer acquires an unpleasant "cardboard"-like taste when stored at a specific temperature and subject to a specified intensity of exposing light. On the contrary, there is another experiment with very many potential pitfalls, which begins with obtaining blood samples from multiple human patients in the operation theater. These are mixed with a saline solution containing an anticlotting agent and a spin trap for off-line EPR analysis in the hope to identify causal relationships between the radical adduct spectra and individual anamneses. In both cases, taking the EPR spectra is easy; the difference is in the complexity of the preceding wet chemistry.

10.2   SPIN LABELS IN ISOTROPIC MEDIA

Spin labels (also called spin probes) are stable $S = 1/2$ paramagnetic radicals. They belong predominantly to the class of nitroxide (also: nitroxyl, aminoxyl) compounds which contain the C-N•-C moiety typically stabilized by three to four methyl groups, for example, $(CH_3)_2$-RC-N•-CR-$(CH_3)_2$. The structure of the prototype spin label TEMPO is shown in Figure 10.3. This basic building block can be modified in many ways to change its aqueous solubility, its two-phase partition coefficient, its bulkiness, its side-chain reactivity, etc. The >N•O radical is a reporter group and, therefore, is not intended to undergo chemical reaction itself. However, single-electron reduction to the diamagnetic hydroxylamine is a not uncommon interfering side-reaction in biological samples as the reduction potential at pH 7 is typically $E_7 \approx +200$ mV (Israeli et al. 2005).

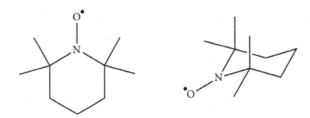

FIGURE 10.3 The structure of TEMPO. The compound 2,2,6,6-tetramethylpiperidine-1-oxyl is a stable radical.

TEMPO, and any of its (not too bulky) derivatives, is comparable in molecular mass with the spin trap DMPO, so the tumbling in water at ambient temperature should again average out all anisotropy. The spectrum is even simpler (namely, three identical lines; see the high-temperature traces in Figure 10.4) than that of the hydroxyl-DMPO adduct because only the ^{14}N nuclear spin contributes to the spectrum:

$$B_{res} = (1/g_{iso})(h\nu/\beta) - A^N_{iso}m^N_I \qquad (10.3)$$

with g_{iso} = 2.0068 and A_{iso} = 16.1 gauss (Moan and Wold 1979). At high resolution also multiple proton hyperfine splittings become apparent (inset to Figure 10.4), and these have been analyzed by (Whisnant et al 1974). In contrast to the result of the spin-trap reaction we have no reaction-efficiency issue because all the spin label is paramagnetic, and so there is usually no detection problem. In fact, the signal-to-noise ratio for, say, 100 μM is so high, that the anisotropic spectrum of a frozen solution (the "powder" spectrum) of the spin label is also easily detected, and its g- and A-tensor principal components are readily determined with Equation 5.11 (assuming tensor colinearity).

Since biomacromolecules tumble so slowly as to afford fully anisotropic EPR spectra in aqueous solutions, a spin label whose motion is severely restricted due to strong interaction with, for example, a protein will also exhibit a "powder" spectrum. This principle is the basis for a spectroscopically straightforward type of experiment in which a chemical group, reactive towards protein, is covalently attached to a spin label. Here is an example for convenience taken from our own work. PCMB (*p*-chloromercuribenzoate) is used to show essentiality of Cys residues for biological activity by abolishing this activity through covalent blocking of the –SH group. The enzyme *p*-hydroxybenzoate hydroxylase from *Pseudomonas fluorescens* has five Cys residues; however, Cys → Ser mutation shows that none of these are essential for activity. On the other hand, PCMB quantitatively reacts with only one Cys resulting in rapid inactivation. When PCMB is covalently attached to a spin label, and this complex is allowed to bind to the protein (PCMB-Cys bond formation) both the wild-type enzyme and all the single Cys→Ser mutants show stoichiometric binding of one label molecule per protein molecule, based on integration of the $S = 1/2$ spectrum, except for the Cys211Ser mutant, whose EPR integrates only to 0.2 spins. This simple 20-minute EPR experiment of recording and integrating six spin-label spectra and a copper standard proves that the PCMB binds exclusively to Cys211.

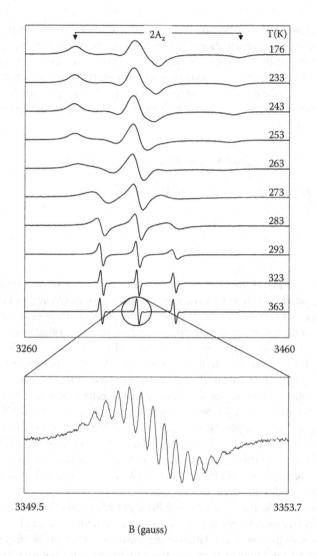

FIGURE 10.4 Anisotropy averaging in the EPR of TEMPO as a function of temperature. The spectra are from a solution of 1 mM TEMPO in water/glycerol (10/90). The "blow-up" of the middle ^{14}N ($I = 1$) hyperfine line in the 90°C spectrum has been separately recorded on a more dilute sample (100 µM) to minimize dipolar broadening and, using a reduced modulation amplitude of 0.05 gauss, to minimize overmodulation. The multiline structure results from hyperfine interaction with several protons.

Combined with structural and activity data, this result can be interpreted in terms of reduced affinity for the aromatic substrate due to the binding of PCMB (van der Bolt et al. 1994).

The goal of the majority of spin-label studies is significantly more ambitious than the simple experiment just described. They are carried out under conditions intermediate between the two extremes of Figure 10.4, that is, between extremely fast

tumbling and no tumbling because they are aimed at obtaining dynamic information, for example, on conformational transitions of proteins, on membrane fluidity, on
membrane permeability and partitioning, etc. The key issue in the analysis of these
experiments is that the paramagnetic properties of individual molecules change on a
time scale that is within the time resolution of the EPR measurement. In terms of the
spin Hamiltonian for an $S = 1/2$ system subject to a hyperfine interaction to a single
nucleus we now have

$$H_S = Tr(g)\beta \boldsymbol{B} \bullet \boldsymbol{S} + \beta \boldsymbol{B} \bullet \boldsymbol{g}'(t) \bullet \boldsymbol{S} + Tr(A)\boldsymbol{S} \bullet \boldsymbol{I} + \boldsymbol{S} \bullet \boldsymbol{A}'(t) \bullet \boldsymbol{I} \qquad (10.4)$$

in which $Tr(M)$ is the trace of a matrix M, that is:

$$Tr(g) = (g_x + g_y + g_z)/3$$
$$Tr(A) = (A_x + A_y + A_z)/3 \qquad (10.5)$$

and the tensors g and A are assumed to be colinear (otherwise, we would write, for
example, A_{xx}, etc.). Equation 10.4 is a way to separate the isotropic from the anisotropic contributions to the spectrum. The traces of the two matrices give the scalar EPR
parameters for the resonance condition in Equation 10.3. The other two terms add the
anisotropic contributions, and together they would result in the "powder" resonance
condition of Equation 5.11 were it not for the "(t)," which indicates that the interactions are time-dependent, that is, they change during the EPR observation due to some
movement of the paramagnet on a time scale that interferes with this observation.

To work out the time-dependence requires a *specific model* for the movement of
the paramagnet, for example, Brownian motion, or lateral diffusion in a membrane,
or axial rotation on a protein, or jumping between two conformers, etc. That theory
is beyond the scope of this book: the math can become quite hairy and can easily fill
another book or two. We limit the treatment here to a few simple approximations that
are frequently used in practice.

What exactly is the time scale of an EPR experiment? In other words, how fast or
slow do molecular movements have to take place in order to have an effect on the EPR
spectrum through the time-modulation of the anisotropic interactions in Equation 10.4?
It is in fact pretty hard to give an exact answer to this question, but we can make order-
of-magnitude estimations. The frequency of an X-band experiment is $v \approx 9.5$ GHz, and
this corresponds to an angular frequency $\omega = 2\pi v \approx 6 \times 10^{10}$ s^{-1}. Therefore, rotational
events that occur on a "typical" timescale of $(6 \times 10^{10})^{-1}$ s or circa 0.02 nanoseconds,
or faster, are "seen" by the spectrometer as fully averaged. And what is the "typical"
timescale of an event? For a molecular rotation we can loosely define this as the average
time required for a molecule to rotate over an angle of, say, one radian. In diffusional
rotation the movement is caused by shearing forces from collisions; the paramagnetic
molecule encounters some other molecule, for example, a solvent molecule, and this can
produce a tangential "kick" of the paramagnet, which sets off to rotate at a speed that
subsequently dies out exponentially as time proceeds. We take the average half life of
this angular-velocity decay as the "typical" time of a rotation to be "seen" in the EPR
experiments, and we call this the rotational correlation time, τ_c. A relatively simple

model for the rotation is the Debye diffusion model through rotational Brownian motion (cf. Carrington and McLachlan 1967: 189) based on the Stokes–Einstein relation:

$$D = kT / (8\pi\eta R^3) \tag{10.6}$$

in which D as the spherical diffusion coefficient, η as the viscosity of the medium, and R as the radius of the molecule modeled as a spherical particle afford:

$$\tau_c = (6D)^{-1} = 4\pi\eta R^3 / (3kT) = \eta V / (kT) \tag{10.7}$$

with V being the volume of the molecule.

In practice, five different time regimes are loosely discerned as defined in Table 10.1. We are already familiar with the two extreme cases: for a small molecule in aqueous solution at ambient temperatures $\tau_c \approx 10^{-12}$ s, that is, the extremely fast regime illustrated by the 363 K spectrum in Figure 10.4. For a frozen solution at cryogenic temperatures $\tau_c \gg 10^{-3}$ s, that is, the rigid regime illustrated by the 176 K spectrum in Figure 10.4. Now let us look at the three intermediate cases.

In the fast regime we still observe an isotropic triplet, but the widths of the individual lines (and, therefore the amplitudes) varies with m_I according to

$$W(m_I) = A + Bm_I + Cm_I^2 \tag{10.8}$$

This expression happens to be identical to Equation 9.26c, but the physical cause of the broadening in the two situations is unrelated (namely: fast rotational diffusion in the liquid state versus A-strain in the solid state). The A-term affects all hyperfine lines equally; the B-term causes a monotonously increasing or decreasing linewidth over the lines, and the C-term makes the outer lines broader than the inner one(s). The A-term is not used to estimate τ_c because it encompasses also inhomogeneous broadening effects not related to rotational diffusion, notably, unresolved hyperfine structure from nuclei other than the nitroxide nitrogen, and instrumental factors, for example, inhomogeneity in the external magnetic field. For nitroxide spin labels the high-field line (which for $A > 0$ has $m_I = -1$) is always found to be the broadest, so B is negative. Furthermore, for

TABLE 10.1
Approximate time regimes in solution EPR

Time regime	Correlation time (s)	EPR spectral phenotype
Extremely fast	$\tau_c < 10^{-11}$	Isotropic multiplet with constant linewidth
Fast	$10^{-11} < \tau_c < 10^{-9}$	Isotropic multiplet with varying linewidth
Slow	$10^{-9} < \tau_c < 10^{-7}$	Spectrum of intermediate shape
Very slow	$10^{-7} < \tau_c < 10^{-3}$	Powder spectrum sensitive to saturation transfer
Rigid	$10^{-3} < \tau_c$	Rigid powder spectrum

nitroxides at X-band, the value of C is typically of the same order of magnitude as that of $|B|$, and so the broadening effect on the $m_1 = 1$ line always stands out.

When the anisotropic spin Hamiltonian parameters g_i and A_i are obtained from the powder spectrum in the rigid limit, one can define the quantities

$$\gamma = (\beta / \hbar)[g_z - (g_x + g_y)/2]$$

$$\alpha = (4\pi / 3)[a_z - (a_x + a_y)/2]$$

(10.9)

in which the a_i are the hyperfine constants in units of MHz, and theory (cf. Nordio 1976) then relates the rotational correlation time τ_c with the linewidth parameters B and C, through γ and α, as

$$B = (4/15)\alpha\gamma B_{res}\tau_c$$

$$C = (1/8)\alpha^2\tau_c$$

(10.10)

An approximate spectral-analysis technique of comparable simplicity has been put forth for the slow time regime of tumbling. The method once more requires additionally taking a true powder spectrum under rigid conditions and is based upon comparison of the "reduced apparent" hyperfine splitting A_z' in the slow-regime spectrum to the true splitting A_z (cf. Figure 10.4), and relating the ratio A_z'/A_z to the rotational correlation time τ_c (Goldman et al. 1972):

$$\tau_c = a[1 - (A_z' / A_z)]^b$$

(10.11)

in which a and b are parameters that depend upon the rotational model, and on the intrinsic linewidth and the value of A_z. For Brownian diffusion and using the simulation data in (Freed 1976: 84) one gets for $A_z = 32$ gauss:

$$a = (2.81 \times 10^{-10})\delta - (0.18 \times 10^{-10})$$

$$b = 0.24\delta - 1.94$$

(10.12)

and scaling to other A_z-values (typically in the range 27–40 gauss for nitroxides) is done by the proportionality

$$S \propto \tau_c A_z$$

(10.13)

The δ in Equation 10.12 is the peak-to-peak derivative Lorentzian linewidth in gauss of the actual spectrum in the slow-tumbling regime, and thus may be difficult to estimate without resorting to simulations, but this value is only slightly larger than that of the isotropic spectrum in the extremely fast regime (Goldman et al. 1972). Consult the cited references (and work quoted therein) for a discussion on the range of τ_c-values for which the simple description, above, is valid and for a full exposé of the theoretical background.

Finally, in the very slow time regime, tumbling has become too slow to affect the regular powder spectrum under nonsaturating conditions, however, when, during a regular scan in which the external magnetic field is slowly scanned, an intermediate B_{res}-position of the powder pattern is partly saturated, then this saturation can be *transferred*

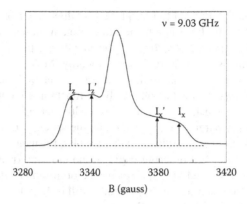

FIGURE 10.5 Estimating τ_c from saturation transfer. In the dispersion spectrum of a spin label (TEMPO) the ratio of I'/I runs from approximately unity in the rigid limit, when the rotation correlation time $\tau_c \approx\geq 10^{-3}$, to approximately zero for $\tau_c \approx 10^{-7}$.

to other parts of the spectrum, leading to specific changes in the overall spectrum (Thomas et al. 1976, Hyde and Dalton 1979). The physics of the phenomenon are based on the rotation correlation time τ_c being comparable to the time T_1 associated with spin-lattice relaxation. The method of saturation-transfer EPR involves a comparison of spectral shape under partially saturating conditions versus nonsaturating conditions, and relating the difference to the rotational correlation time τ_c. The technique is based on the fact that saturation transfer is much more efficient at intermediate field positions of the powder pattern (that is, at intermediate orientations away from the canonical orientation that produce the extreme spectral turning points along the x-and z-axis of the molecular axes system) than at the extreme positions. The effect is not easily measured on the regular EPR spectrum, but some spectrometers have an option to change from absorption-derivative to dispersion-derivative measurement detected 90° out of phase with respect to the 100 kHz field modulation (this is effected by a simple switch of a knob on the microwave bridge), which for randomly oriented samples, results in a negative absorption-shaped powder pattern, and this is rather sensitive to saturation transfer as illustrated in Figure 10.5: the ratio of intensity at the indicated intermediate position over that at an extreme position, I_x'/I_x or I_z'/I_z, changes approximately linearly from its rigid-limit value at $\tau_c \geq 10^{-3}$ to approximately zero at $\tau_c \approx 10^{-7}$. Note that the determination can be semi-quantitative in the $10^{-6} < \tau_c < 10^{-4}$ range and is much less reliable in the two flanking decades. The signal-to-noise ratio of the experiment can be improved by detecting the absorption-derivative detected 90° out of phase with respect to the second harmonic of a 50 kHz field modulation, but this requires the construction of an extension to the electronics of the spectrometer (ibidem).

10.3 SPIN LABELS IN ANISOTROPIC MEDIA

Note that in the entire discussion on spin labels, above, we have assumed *isotropic* rotational diffusion. In several real situations of biochemical relevance this is not a tenable assumption. Perhaps the most significant one occurs when the nitroxide spin label is covalently attached to phospholipids which are subsequently built into membranes, or when the label is attached to a protein immersed in a membrane. Associated with the specific structure of the membrane phospholipid bilayer (in Archaea: monolayer) is a nonisotropic directionality. Molecules in the membrane tend to align along an axis normal to the membrane plane (parallel to the chain axis of the lipid); this tendency is not a constant but a function of the flexibility, or fluidity, of the membrane, which, for one, increases with increasing temperature, but it also depends on other parameters, for example, the specific structure of the lipids. In other words, depending on a collection of variables, the membrane is more or less ordered in a range from full ordering (rigid) to no ordering (isotropic fluid). Such a system is usually described by a macrovariable or an *order parameter*, S, whose magnitude ranges from zero (no order) to unity (full order): $0 \leq S \leq 1$. For example, when an individual axial lipid molecule in a membrane makes an angle θ with the "director" (i.e., the tendency to align along the normal), the state of the membrane is defined by its order parameter

$$S = \left\langle \frac{3\cos^2\theta - 1}{2} \right\rangle \tag{10.14}$$

in which the $\langle \rangle$ indicate averaging over many molecules. This order parameter, and therefore the degree of fluidity of the membranous system, can be determined from the EPR spectrum of an "axial" spin label (e.g., a nitroxide attached to a lipid) by comparison of its spectrum in the fluid membrane, with hyperfine splittings A', to its rigid powder pattern, with splittings A (Hubbell and McConnell 1971):

$$S = \frac{A_\parallel' - A_\perp'}{A_z - A_x} \frac{a_N}{a_N'}$$

$$a_N = (1/3)(A_x + A_y + A_z) \tag{10.15}$$

$$a_N' = (1/3)(A_\parallel' + 2A_\perp')$$

or in a slightly simplified form (Gaffney and McConnell 1974):

$$S = \frac{A_\parallel' + A_\perp'}{A_z - (A_x + A_y)/2} \tag{10.16}$$

In recent years spin label EPR has received a boost due to developments in molecular biological techniques, which now make it practical to systematically replace essentially all amino-acids in protein sequences by cysteines through systematic side-directed mutagenesis (SDM) (Hubbell et al. 1998). Employing spin labels covalently modified with a group that has specific reactivity towards the thiolato side group of cysteine, each amino acid can be subjected to side-directed spin labeling (SDSL) and then studied with SDSL-EPR. With this approach two new determinations become possible: (1) global mapping of flexibility and accessibility and (2) specific 3-D structure from distance constraints of pairs of labeled cysteines (Oda et al. 2003; Lagerstedt et al. 2007).

Accessibility is qualitatively determined by addition to the solution of a faster relaxing paramagnet, for example, Cr^{III}oxalate, which will broaden the nitroxide spectrum when it is at a short distance. Similarly, close spatial distance between two labeled Cys residues is deduced from mutual spectral broadening when compared to the sum of their individual spectra from singly-labeled protein. Quantitative determination of distances and relative orientation between two labels is in principle possible by detailed analysis of the dipolar interaction (see Chapter 11). Finally, note that the experiment is similar to that of high-resolution NMR in solution, but only conceptually so: global labeling (cf. ^{15}N labeling for NMR) is, of course, impossible, because the result would be poly-Cys. On the other hand, specific labeling by construction of a double (or a multiple) Cys mutant does not have an equivalent in NMR, and it results in relatively straightforwardly analyzable EPR.

10.4 METALLOPROTEINS IN SOLUTION

In metalloproteins, the paramagnet is an inseparable part of the native biomacromolecule, and so anisotropy in the metal EPR is not averaged away in aqueous solution at ambient temperatures. This opens the way to study metalloprotein EPR under conditions that would seem to approach those of the physiology of the cell more closely than when using frozen aqueous solutions. Still the number of papers describing metalloprotein bioEPR studies in the frozen state by far outnumbers studies in the liquid state. Several additional theoretical and practical problems are related to the latter: (1) increased spin-lattice relaxation rate, (2) (bio)chemical reactivity, (3) unfavorable Boltzmann distributions, (4) limited tumbling rates, and (5) undefined g-strain.

Most biochemically relevant high-spin systems have such short T_1-relaxation times that their EPR is broadened beyond detection at ambient temperatures. An exception is the class of $S = 5/2$ Mn^{II} systems with $D \ll h\nu$. Also, $S = 7/2$ Gd^{III}-based MRI shift reagents exhibit readily detectable room-temperature EPR spectra. Otherwise, aqueous-solution transition ion bioEPR is limited to complexes of $S = 1/2$ metals, in particular, Cu^{II}, and to a lesser extent $V^{IV}O^{2+}$, Ni^{III}, Ni^{I}, Mo^{V}, and W^{V}. Cupric is the stable oxidation state of biological copper under aerobic conditions, however, the other metals are stable as V^{V}, Ni^{II}, Mo^{VI}, and W^{VI}, and, therefore, the other oxidation states associated with $S = 1/2$ paramagnetism may exhibit oxidative or reductive reactivity and may thus require specific experimental precautions such as strict anaerobicity over the course of the EPR experiment.

Figure 10.6 shows the $I = 3/2$ parallel hyperfine pattern of the $S = 1/2$ cupric site in the mammalian CuZn enzyme superoxide dismutase. The top trace is from a frozen aqueous solution, the middle trace is from a 10% glycerol containing frozen aqueous solution, and the bottom trace is from an aqueous solution at room temperature. Addition of glycerol causes small but significant spectral changes that indicate contributions from g-strain and A-strain to the spectrum presumably reflecting the influence of hydrostatic stress associated with ice crystal formation. The spectrum taken from the liquid-phase sample exhibits somewhat more pronounced changes, notably a reduction in the apparent A_z-value and changes in linewidths of the individual hyperfine peaks. It is not clear whether these changes are caused by modified strain or by partial averaging ($\tau_c \approx 10^{-8}$) of the anisotropic A-tensor, or both. Furthermore,

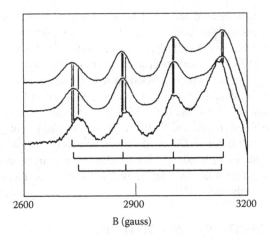

2600   2900   3200

B (gauss)

FIGURE 10.6 Comparison of solid-state and liquid-state spectra from a copper protein. The figure illustrates shifts in apparent g_z and A_z-values of the $S = 1/2$ and $I = 3/2$ spectrum from Cu^{II} in bovine superoxide dismutase as a function of the surrounding medium. Top trace: frozen aqueous solution at 60 K; middle trace: frozen water/glycerol (90/10) solution at 60 K; bottom trace: aqueous solution at room temperature. (Modified from Hagen 1981.)

the "high" temperature appears to cause some homogeneous contribution (i.e., lifetime broadening) to the overall linewidth. Finally, the optimal sample temperature for X-band EPR of mononuclear cupric proteins is typically circa 50–60 K, so even in the absence of lifetime broadening an increase to circa 300 K implies a decrease of signal-to-noise ratio by a factor of five by virtue of Curie's law.

11 Interactions

Metalloproteins frequently contain more than one paramagnetic center. The EPR of such systems is not simply the sum of the spectra of the individual centers. The different paramagnets influence each other through spin–spin interaction just like the unpaired electrons in a high-spin system ($S*S$), but now the interaction is between *different* spins (S_1*S_2) and it is furthermore not necessarily limited to two centers. The two key parameters in this game are *distance* and relative *orientation*. The distance may be as large as several nanometers (i.e., many chemical bonds) when the centers are located in different parts of the protein, or even in two different proteins, but it can also be as small as a few angströms (i.e., one or two chemical bonds) when paramagnetic metal ions form a dinuclear or a polynuclear cluster. The microwave frequency plays a similar role as for isolated high-spin systems: spin–spin interactions are independent of the external magnetic field, so an increase in the frequency/field couple, v/B, makes the relative contribution of interactions to the EPR spectrum less important compared to the electronic Zeeman interaction of the individual centers. Eventually the spectral effects of interaction become insignificant; for weak interactions this may occur at frequencies in or slightly above X-band, but for strong interactions this may not be reachable for any practical EPR frequency.

Interactions between two different spin systems come in two kinds: dipole–dipole interactions and exchange interactions. Dipolar interactions are purely magnetic in nature, are significant over long distances, and are relatively weak. Exchange interactions are electrostatic in nature, are important only over short distances, and can be orders of magnitude stronger than dipolar interactions. Dipolar interactions are operative "through space" (including vacuum); exchange interactions work "through bond," that is, directly between two atoms (through a single bond) or indirectly via intermediate atoms (through multiple bonds).

Analysis of biomolecular EPR spectra with interaction can be complicated; the number of formally required parameters can be so large as to preclude finding a unique solution. The goal of this chapter is to learn how to read interaction spectra in a semiquantitative manner, at best, and to be able to decide what information can be extracted without laborious in-depth analysis, or even to make the interaction altogether disappear by simple physical-chemical means.

11.1 DIPOLE–DIPOLE INTERACTIONS

The classical interaction energy between two point dipolar magnets with moments μ_a and μ_b is

$$E_{ab} = \frac{\mu_a \bullet \mu_b}{r^3} - \frac{3(\mu_a \bullet r)(\mu_b \bullet r)}{r^5} \tag{11.1}$$

in which r is the vector that joins the (center of gravity of) the two vectors μ_i. Note the importance of the distance, r, and the relative orientation, that is, the angle θ between the two vectors μ_a and μ_b (from the definition of the dot product: $\boldsymbol{a} \bullet \boldsymbol{b} = |\boldsymbol{a}||\boldsymbol{b}|\cos\theta$), which is readily illustrated with two toy bar magnets (as an approximation of point dipoles). When bringing them together in parallel orientation we experience a rapidly increasing repulsion ($E > 0$) and in antiparallel orientation a rapidly increasing attraction ($E < 0$). There is also a "magic angle" effect: when the bars are antiparallel and the joining vector makes an angle $\theta \approx 54.74$ degrees, then $E \approx 0$, that is, the dipolar interaction vanishes.

The quantum-mechanical equivalent of Equation 11.1 for two paramagnets with magnetic moments $\mu_i = \beta \boldsymbol{g}_i \bullet \boldsymbol{S}_i$ is

$$H_{dip} = \frac{\mu_0 \beta^2}{4\pi r^3}\left[(g_a \bullet S_a) \bullet (g_b \bullet S_b) - \frac{3(r \bullet g_a \bullet S_a)(r \bullet g_b \bullet S_b)}{r^2} \right] \tag{11.2}$$

and we can write the complete spin Hamiltonian for such a system as the sum of the individual electronic Zeeman interactions plus the joint dipole–dipole interactions as

$$H_S = \beta B \bullet (g_a \bullet S_a + g_b S_b) + H_{dip} \tag{11.3}$$

In the early nonbiological EPR literature, this theme has been firmly associated with light-excited, phosphorescent organic biradicals such as the naphthalene molecule. When these diamagnets are kept at low temperature, say 77 K, and irradiated with a UV-lamp, an electron is excited and a biradical is created, which relatively slowly ($\tau_{0.5}$ is circa 3 s for naphthalene) falls back to the ground state by phosphorescence. Upon continuous illumination a significant steady-state concentration of biradical is created sufficient for triplet EPR detection (De Groot and Van der Waals 1960; Wasserman et al. 1964). The excited electron originated from a Pauli pair, and it leaves behind a now unpaired electron, so we end up with two unpaired electrons which are in fact indistinguishable: an electronic triplet ($S = 1$) has been created. $S = 1$ means that we consider the spins, S_a and S_b, of the two electrons to have merged into a new entity with a system spin S_{ab}, or S for short. This is, indeed, the approach usually taken for the description and analysis of EPR from biradicals, namely the spin Hamiltonian:

$$H_S = \beta B \bullet g \bullet S + S \bullet D \bullet S \tag{11.4}$$

and spin wavefunctions

$$\varphi_i = |m_S\rangle = \{|+1\rangle; |0\rangle; |-1\rangle\} \tag{11.5}$$

with possible complications from tensor noncolinearity as discussed in Section 8.2. In fact, the g-tensor for these organic biradicals is essentially isotropic ($g = 2.0028$ for naphthalene triplet), and, recalling the traceless nature of the D-tensor (cf. Equation 7.34), we can write

$$H_S = g\beta B S_z + D[S_z^2 - S(S+1)/3] + E(S_x^2 - S_y^2) \tag{11.6}$$

The $S*S$ term describes dipolar interaction: the elements of the D-tensor in Equation 11.4 are directly related to the dipole–dipole interaction by averages over the electronic wavefunction:

$$D_{xx} = \frac{\mu_0 (g\beta)^2}{8\pi} \left\langle \frac{r_{ab}^2 - 3x_{ab}^2}{r_{ab}^5} \right\rangle; \ D_{xy} = \frac{\mu_0 (g\beta)^2}{8\pi} \left\langle \frac{-3x_{ab}y_{ab}}{r_{ab}^5} \right\rangle; \ \text{etc.} \qquad (11.7)$$

In the excited naphthalene molecule the two unpaired electrons are spatially confined due to the limited size of the molecule.

In metalloproteins two paramagnets can be much farther apart, and so the dipolar interaction can be correspondingly weaker. Furthermore, the centers will usually each have significant g-anisotropy, and their local structures will differ and will have a complex mutual geometrical relationship. We therefore use the symmetric biradical as a simple model to obtain a first impression of the type of spectral patterns to be encountered.

Figure 11.1 shows a series of calculated triplet X-band ($h\nu \approx 0.3$ cm^{-1}) spectra for two identical, isotropic $S = 1/2$ systems and colinear g- and D-tensors. The axial zero-field splitting parameter ranges from 0 to 0.05 cm^{-1} in steps of 0.01 cm^{-1} and the rhombicity $\eta = E/D = 0.1$. Since the axial splitting in zero field ΔE_{ZF} (i.e., the splitting between the $|\pm1\rangle$ non-Kramer's pair and the $|0\rangle$ level) is identically equal to D for $S = 1$ (cf. Chapter 5, Figure 5.16), we have $\Delta E_{ZF} < h\nu$, and there are three possible transitions, namely, two allowed ones and a forbidden transition

$$|\pm1\rangle \leftrightarrow |0\rangle \ ; \quad |\Delta m_S| = 1$$
$$|+1\rangle \leftrightarrow |-1\rangle \ ; \quad |\Delta m_S| = 2 \qquad (11.8)$$

The label $|\Delta m_S| = 2$ does *not* mean that two quanta $h\nu$ are absorbed; it is simply a somewhat unfortunate but widely divulged notation to indicate a transition ($\Delta E = h\nu$) between two levels that we happen to have labeled as $|+1\rangle$ and $|-1\rangle$. Strictly speaking, these labels should only apply to the strong-field situation of ($S*S$) $\ll$ ($B*S$) as we discussed in Chapter 4. In the present example of Figure 11.1 we are in the weak-to-intermediate field regime ($S*S \gtrsim B*S$), which means that the actual wavefunctions are linear combinations of the ones in Equation 11.5. In particular, a rhombic E-term mixes the $|+1\rangle$ and $|-1\rangle$ levels as can be seen from its appearance in nondiagonal positions in the zero-field energy matrix

$$
\begin{array}{l}
\langle+1| \\
\langle-1| \\
\langle0|
\end{array}
\begin{vmatrix}
D/3 & E & 0 \\
E & D/3 & 0 \\
0 & 0 & -2D/3
\end{vmatrix}
\qquad (11.9)
$$

The "forbidden" $|\Delta m_S| = 2$ transition occurs between levels that separate in an increasing magnetic field twice as rapidly as the levels of the "regular" $|\Delta m_S| = 1$ transitions, and so this transition is expected to be found at "half field," that is, at a field corresponding to an effective g-value $g_{eff} = 2 \times g \approx 4$. In Figure 11.1 we see the relative intensity

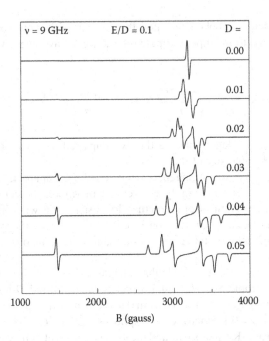

FIGURE 11.1 Schematic triplet spectra as a function of zero-field splitting. The X-band $S = 1$ spectra have been calculated for indicated D-values and with $E = D/10$ to illustrate increasing intensity of the $|\Delta m_S| = 2$ transition at half field with increasing D-value.

of the forbidden transition to increase with respect to those of the allowed transitions as a function of increasing zero-field interaction strength, and the intensities become comparable for $D \approx 0.05$ cm^{-1}. Naphthalene biradical has an even greater zero-field interaction ($D = 0.1$ cm^{-1}, $E = 0.015$ cm^{-1}) and the half-field transition has become the dominant feature of the spectrum (Wasserman et al. 1964; Weltner 1989).

11.2 DIPOLAR INTERACTION IN MULTICENTER PROTEINS

For two centers in a metalloprotein the mutual dipole–dipole interaction is usually weaker, and the intensity of the half-field line is typically one to three orders of magnitude less than that of the allowed ones. The canonical example is that of the eight-iron ferredoxin, that is, a relatively small protein of circa 9 kDa, carrying two iron–sulfur cubanes, which in the reduced protein both are [4Fe-4S]$^{1+}$ and $S = 1/2$. Magnetically isolated reduced cubanes typically exhibit a simple, rhombic $S = 1/2$ spectrum with approximate g-values $g_z \approx 2.03$–2.10 and $g_{xy} \approx 1.96$–1.85. For the 8Fe ferredoxin the main spectral features are also in this g-value range, but the simple rhombic pattern is no longer recognizable as it is "deformed" by dipole–dipole interaction resulting in shifts and splittings of the individual $S = 1/2$ powder shapes. Furthermore, a weak, but usually distinct $|\Delta m_S| = 2$ feature can be detected at half field when using a high spectrometer gain setting combined with relatively high microwave power levels (Mathews et al. 1974). The iron ions of the 8Fe ferredoxin

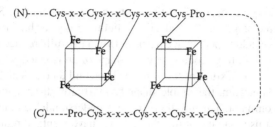

FIGURE 11.2 Common cysteine binding pattern for two interacting cubanes.

are coordinated by the thiolato side groups of eight Cys residues with a peculiar, intertwined sequence pattern (depicted in Figure 11.2) that contributes to the relative closeness of the two cubanes in space. This is an extremely common binding motif in numerous proteins including rather complex ones (Figure 7.4), and so it is worthwhile to make a mental note of its approximate shape for recognition purposes. The next four figures illustrate the two-cubane spectral pattern with data taken from two complex enzymes, and they also indicate two "escape routes" to turn the irregular patterns into simple, isolated cubane spectra.

Figure 11.3 shows spectra from a cubane pair in an enzyme called DPD, dihydropyrimidine dehydrogenase (Hagen et al. 2000). It catalyzes the first step in the breakdown of pyrimidine bases. The two cubanes have unusually low reduction potentials, and so they are reduced to the [4Fe-4S]$^{1+}$ form with $S = 1/2$ ground state by means of

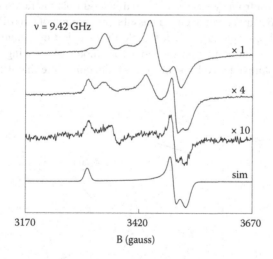

FIGURE 11.3 Spectral changes due to increased coupling between two cubanes. A pair of cubanes in the porcine enzyme dihydropyrimidine dehydrogenase is increasingly (from bottom to top trace) reductively titrated. Initially, a single, magnetically isolated [4Fe-4S]$^{1+}$ spectrum is found (simulated in the bottom trace); at intermediate degree of reduction an overlap of this spectrum with that from interacting cubanes is observed; at high degree of reduction only an interaction signal is found. Amplitudes have been rescaled to emphasize changes in spectral shape.

the so-called "deazaflavin/light" method, which was explained in Section 3.6. The figure shows three stages in the reduction: slightly reduced, half reduced (i.e., one electron equivalent added per cubane pair), and virtually fully reduced. The first spectrum is the simple, slightly rhombic powder shape of an isolated $S = 1/2$ cubane; the second spectrum is a mixture of this simple shape and a complex "interaction" pattern; in the last spectrum, the simple shape has virtually disappeared and the complex pattern is dominant. This titration course is understandable in terms of a simple redox model. Let us assume that the two cubanes have identical reduction potentials and, furthermore, that their individual rhombic EPR spectra are also identical. Then, adding increasing amounts of reducing equivalents (from 0–2 electrons) will create varying populations of protein molecules with either two oxidized cubanes, or one of the cubanes reduced, or both of them reduced. For example, when we add exactly one reducing equivalent per protein molecule, then the probability to find a protein molecule with cubane sites A and B in one of the four possible configurations, $A_{ox}B_{ox}$, $A_{red}B_{ox}$, $A_{ox}B_{red}$, $A_{red}B_{red}$, is one quarter. The fully oxidized form has a diamagnetic ground state, so with one reducing equivalent added, the relative spin count (integrated intensity) for the "simple" spectrum over the "interaction" spectrum is 2.

In Figure 11.4 the fractional populations of the different EPR detectable redox forms under this simple model are plotted as a function of added reducing equivalents from a strong reductant. Real situations are likely to be more complicated because the two cubanes have different g-values and/or reduction potentials, but the general message of Figures 11.3 and 11.4 is that it is worthwhile considering going through the trouble of setting up some form of redox titration for a better understanding of spectra from systems subject to dipole–dipole interaction. In its simplest form such an experiment would encompass the preparation of two EPR samples only (or even two subsequent states for a sample in a single tube), namely, a partially-reduced sample by addition of substoichiometric amount of reducing equivalents, and a fully-reduced sample by addition of excess reductant. Note that the best resolution

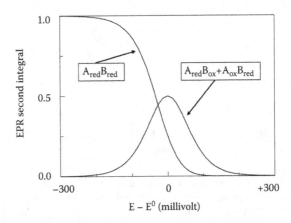

FIGURE 11.4 Fractional populations of half- and fully reduced pairs. The traces are for the idealized case of two $S = 1/2$ systems with identical EPR spectrum and identical reduction potential $E^0 = 0.0$ V.

of individual spectral components is obtained for low degrees of reduction, but the quality of such an experiment is of course counterbalanced by low signal intensity.

A common mistake (or call it a "first approximation") in the analysis of these types of spectra is to deconvolute them as a simple sum of noninteracting species, and thus to ignore interaction altogether. The outcome of such an analysis is likely to be the identification of a higher number of individual spectral components than present in actuality, and this in turn leads to biochemically incorrect conclusions about substoichiometry and/or inhomogeneity of prosthetic groups. Some authors use the technique of partial saturation for the deconvolution based on the assumption that individual spectra may respond differently to microwave power saturation. Having read through Chapter 9, it should be clear that this is not a good idea at all, because spectra of systems subject to g-strain (such as $S = 1/2$ cubanes) have power-saturation characteristics that are not constant over a single, individual powder pattern.

Recognition of the occurrence of interaction is clearly of importance, and in addition to the redox titration approach, we can do two more experiments to identify dipole–dipole interaction. The first one is to look for the "forbidden" transition, which is not always trivial because the relative weakness of dipolar interactions between centers in metalloproteins typically affords only low intensities. It helps that we know approximately where to find the transition, namely at a resonance field corresponding to circa $2 \times g$. Figure 11.5 is the half-field spectrum corresponding to the maximally reduced cubane pair in Figure 11.3. A frequent practical complication can be the fact that $2g$ is close to the effective $g \approx 4.3$ from the ubiquitously present "dirty iron" contaminant (cf. Chapter 5), and the latter signal may be interfering to the extent that the half-field spectrum may not be recognizable anymore.

Another experiment to recognize interaction is based on its *in*dependence of the microwave frequency. If we increase the frequency, then the Zeeman interaction will gain relative importance, and the shape of the spectra should simplify. Experimentally, this may turn out to be a difficult approach due to the rapidly

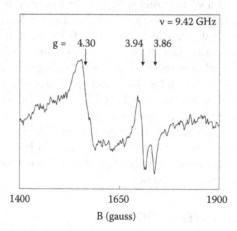

FIGURE 11.5 Half-field spectrum of two interacting cubanes. The signal from dihydropyrimidine dehydrogenase is of low intensity due to the relative weakness of the dipolar interaction. The $g = 4.3$ signal is a "dirty iron" contamination.

decreasing signal-to-noise ratio with increasing frequency for, for example, iron–sulfur proteins (cf. Priem et al. 2005).

Analysis of the spatial and distance information in dipole–dipole interaction would appear to be a potentially interesting method for biomacromolecular structure determination as an alternative to x-ray crystallography or high-resolution multidimensional NMR spectroscopy. A *qualitative* example was alluded to in the previous chapter, namely, the outlining of global folding of a flexible C-terminal domain of the human apolipoprotein A-1 (a structural component of HDL: high density lipoprotein complex) by observation of mutual EPR broadening of spatially close, but sequentially distant, pairs of spin-labeled Cys residues (Oda et al. 2003). However, prospects for *quantitative* applications to systems of otherwise unknown structure are rather more dim at this time.

The first serious attempt in this direction was the analysis of multi-frequency (3–15 GHz) EPR data from the enzyme TMAdh (trimethylamine dehydrogenase), a protein that, upon two-electron reduction by its substrate trimethylamine, exhibits triplet spectra resulting from the interaction between an $S = 1/2$ [4Fe-4S]$^{1+}$ cubane and an $S = 1/2$ FMN (flavin mononucleotide) semiquinone radical (Stevenson et al. 1986). The center-to-center distance deduced from the EPR analysis came out as circa 3–5 Å, but a subsequent x-ray crystallographic structure determination to 2.4 Å resolution afforded a distance of circa 12 Å (Lim et al. 1986). In a later study with multi-frequency EPR in the extended range of 9–340 GHz it was concluded that reliable distance information cannot be extracted because the point-dipole model (intrinsic in Equations 11.1 and 11.2) is an unacceptable simplification for centers whose spins are significantly delocalized over a spatial structure (here: a cubane with significant spin density on all four irons and a three-ring heteroaromate with a delocalized spin) (Fournel et al. 1998). Furthermore, the two centers turn out to be also subject to substantial exchange interaction (see below).

In sharp contrast to this case, in a similar system of an $S = 1/2$ FMN radical and an $S = 1/2$ [2Fe-2S]$^{1+}$ dinuclear cluster with a center-to-center distance of circa 12 Å in the enzyme PDR (*phthalate* dioxygenase reductase) the X-band EPR showed no evidence whatsoever of dipole–dipole coupling (Bertrand et al. 1995). With a 2 Å resolution x-ray structure available (Correl et al. 1992) this could be explained by the particular mutual spatial orientation of the two paramagnets resulting in an angle between the magnetic vectors of the flavin and the cluster of circa 131° (Bertrand et al. 1995), which is close to the magic angle of 54.7° (or 180-54.7) in which the interaction vanishes in the point-dipole approximation.

The bottom line: quantitative distance data are hard to get from dipolar interaction data, but qualitative or topological information can be obtained. It is usually helpful to study spectra from intermediate redox-titration samples and/or spectra taken at different microwave frequencies.

11.3 EXCHANGE INTERACTIONS

From Equations 11.1 and 11.2 we have seen that the strength of the dipole–dipole interaction decreases rapidly with increasing distance between two paramagnetic centers, and still we choose to call this a *long-range* interaction. The justification for this at-first-sight contradictory attribute is that when the distance, r, between two paramagnets is decreased, another type of interaction, of completely different physical nature, takes over in importance, once we squeeze down r to a few ångströms, that is, a *short-range* interaction. The buzzword here is "exchange," and just like with any buzzword we should be on guard for a suggestion of well-defined content where in fact the word may mean rather different things to different people.

The basic concept is simple enough: When two paramagnets are within a mutual distance of one, or a few chemical bonds, then their electronic wavefunctions will have finite overlap, and so their individual paramagnetism will be mutually influenced. Note that this is a "through-bond" interaction in contrast to the dipolar "through-space" interaction. An obvious question would then be, through what bond? And in the context of biomolecular paramagnetism the embarrassing answer would have to be that we don't really know for sure. Is there such a thing as metal–metal bonds in biology, or is the only through-bond communication via (one or more) bridging ligands, or is reality intermediate between these two extremes? In the earlier physics literature on transition ions in diamagnetic host crystals, the two extreme possibilities had been labeled direct exchange versus superexchange or indirect exchange. The spin Hamiltonian formalism to describe direct exchange is identical to that for superexchange, namely, for two centers with spins S_A and S_B (Moriya 1960; Anderson 1963; Stevens 1997)

$$H_{exchange} = C2JS_A \bullet S_B \quad \text{(isotropic exchange or Heisenberg interaction)}$$

$$+D \bullet S_A \times S_B \quad \text{(asymmetric exchange or Moriya–Dzialoshinski interaction)}$$

$$+S_A \bullet K \bullet S_B \quad \text{(anisotropic exchange or pseudo-dipolar interaction)}$$

$$(11.10)$$

The first term is characterized by a scalar, J, and it is the dominant term. Be aware of a convention disagreement in the definition of this term: instead of $-2J$, some authors write $-J$, or J, or $2J$, and a mistake in sign definition will turn the whole scheme of spin levels upside down (see below). The second and third term are induced by anisotropic spin-orbit coupling, and their weight is predicted to be of order $\Delta g/g_e$ and $(\Delta g/g_e)^2$, respectively (Moriya 1960), when Δg is the (anisotropic) deviation from the free electron g-value. The D in the second term has nothing to do with the familiar axial zero-field splitting parameter D, but it is a vector parameter, and the "×" means "take the cross" product (or vector product); an alternative way of writing is the determinant form

$$H_{asymmetric} = \begin{vmatrix} D_1 & D_2 & D_3 \\ S_{A_1} & S_{A_2} & S_{A_3} \\ S_{B_1} & S_{B_2} & S_{B_3} \end{vmatrix} \quad (11.11)$$

which can be written out as

$$H_{asym} = D_1(S_{A_2}S_{B_3} - S_{A_3}S_{B_2}) + D_2(S_{A_3}S_{B_1} - S_{A_1}S_{B_3}) + D_3(S_{A_1}S_{B_2} - S_{A_2}S_{B_1}) \quad (11.12)$$

Note that the subscript axes indices are 1, 2, 3, and not x, y, z, to indicate that the term is diagonal in an axis system that is generally different from that which diagonalizes the g-tensor of system A and/or B. This means that a full characterization of the asymmetric exchange does not only require values for the three elements of the D-vector, but also three angles over which to rotate to get to the g-diagonalizing reference frame. For example, if we would take the latter to refer to the g-tensor of center A, then we should in general write

$$H_S = \beta B \bullet (g_A \bullet S_A + Q^{-1} \bullet g_B \bullet Q \bullet S_B) + R^{-1} \bullet D \bullet R \bullet (S_A \times S_B) + S_A \bullet R^{-1} \bullet K \bullet R \bullet S_B$$

$$(11.13)$$

in which matrix Q rotates the g_B-tensor to the g_A reference system, and R is the rotation matrix to get from the 1,2,3 coordinates to the g_A system. Furthermore, note that the last term is sometimes called the pseudo-dipolar interaction for the simple reason that it has mathematically the same form as the true dipole–dipole interactions, which creates a problem of how to disentangle the two.

The intimidating hairiness of Equation 10.13 (for a two-center system: 21 parameters plus their possible distributions) is reflected in the paucity of biomolecular examples in which its application has been tried. Characteristically, the asymmetric exchange has generally been ignored in bioEPR, even though it is expected to be more important than the anisotropic term especially since its canceling due to inversion symmetry (ibidem) is pretty unlikely in biomolecules. In a few cases a general zero-field interaction has been used in the analysis of bioEPR to cover both anisotropic exchange and dipole–dipole interaction in a single term. A case in point is the interaction spectra from the enzyme TMAdh that we discussed above in the frame of dipolar interaction. Their analysis afforded a value of $J = -0.36$ cm^{-1} for the isotropic exchange parameter (Fournel et al. 1998), which implies that dipole–dipole interaction is not the only relevant term. Furthermore, it raises the question whether such a complex should not be considered as a single, fully entangled spin system rather than as a collection of individual, weakly interacting paramagnets. In point of fact, a $-2J$ ≈ 0.7 cm^{-1} may not be a very strong exchange coupling, but it is more than two times greater than the X-band microwave quantum $h\nu \approx 0.3$ cm^{-1}. A practical approach for such a system would be to effectively describe the X-band spectrum as that of an $S =$ 1 system (parallel coupled spins) with the spin Hamiltonian

$$H_S = \beta B \bullet g_{AB} \bullet S_{AB} + S_{AB} \bullet D_{AB} \bullet S_{AB}$$
$$(11.14)$$

where for simplicity (or as a first approximation) we have assumed colinearity of the tensors, and in which it is to be understood that g_{AB} is a system g-tensor containing contributions from the g-tensors of the individual spin systems, and the D_{AB} tensor collects dipolar effects, asymmetric and anisotropic exchange effects, and possibly individual zero-field interaction effects (for systems with individual $S > 1/2$). The only practical difference with a bona fide isolated $S = 1$ system (e.g., Ni^{2+}) is then the occurrence of an $S_{AB} = 0$ state (antiparallel coupled spins) at $E = -2J$, which slightly complicates the temperature dependence of the overall signal intensity due to Boltzmann (de)population. For this particular case of $-2J = 0.7$ cm^{-1} the two states ($S = 1$ and $S = 0$) are both circa 50% populated except at very low temperatures (for $T \to 0$ K the $S = 1$ state goes to 100% population).

For relatively simple systems of high symmetry (or for systems assumed to be simple) the system spin Hamiltonian parameters are readily relatable to those of the individual centers, for example,

$$g_{AB} = k_A g_A + k_B g_B \tag{11.15}$$

with

$$k_A = \frac{S_{AB}(S_{AB}+1) + S_A(S_A+1) - S_B(S_B+1)}{2S_{AB}(S_{AB}+1)} \tag{11.16}$$

and a mirrored expression for k_B. For two $S = 1/2$ systems, this gives $k_A = k_B = \frac{1}{2}$, and for the TMAdh example with a rhombic g-tensor for the $[2Fe\text{-}2S]^{1+}$ cluster, and an isotropic g-value for the FMA radical, the result is

$$g_{i(i=x,y,z)} = \left(g_i^{Fe/S} + g_{iso}^{FMN}\right)/2 \tag{11.17}$$

The [2Fe-2S] dinuclear cluster itself is a classical example of such a system paramagnet: in the reduced (1+) state it is made up of a high-spin ferric ion ($S = 5/2$) and a high-spin ferrous ion ($S = 2$), coupled antiparallel into a system spin $S = 1/2$. Under the simple model of Equations 11.15 and 11.16 the system g-values are (Gibson et al. 1966)

$$g_i = (7/3)g_i^{ferric} - (4/3)g_i^{ferrous} \tag{11.18}$$

With a very simple model for the individual iron ions, namely, an isotropic $g = 2.019$ for the Fe^{III} and an anisotropic g-tensor for Fe^{II} in tetragonal symmetry

$$g_x^{ferrous} = g_e + 6\lambda/\Delta_{yz}$$
$$g_y^{ferrous} = g_e + 6\lambda/\Delta_{zx} \tag{11.19}$$
$$g_z^{ferrous} = g_e$$

in which λ is the ferrous ion spin orbit coupling constant and Δ is the crystal field splitting between the $|z^2\rangle$ single d-electron ground state and the $|ij\rangle$ excited state, the Gibson model in Equation 11.19 predicts two of the cluster g-values to be significantly less than g_e, and the third cluster g-value to be greater than g_e, a prediction that was observed experimentally (cf. Figure 9.2) and until 1966 vigorously contested to be a possible EPR spectrum for an iron coordination complex.

The very first EPR pattern ever analyzed in terms of exchange interaction is the X-band spectrum of a crystal of copper acetate hydrate (Bleaney and Bowers 1952), whose stoichiometry is written as $Cu_2(CH_3COO)_4 \bullet 2H_2O$ because it forms dimers of Cu(II) bridged by four carboxylato ligands and with a water molecule at each end of the dimer (van Niekerk and Schoening 1953). The powder of copper acetate is a cheap, stable, and easy to measure example compound for exchange interaction.

The copper dimer is subject to a very strong isotropic (super)exchange interaction with $2J \approx 300$ cm^{-1} (Bleaney and Bowers 1952, Elmali 2000) resulting in an $S = 0$ ground state (antiparallel coupling of the two $S = 1/2$ spins) and an $S = 1$ excited state (parallel coupling). This affords a triplet spectrum with maximal intensity near ambient temperatures, which, upon cooling, sharpens up due to reduction of the spin-lattice relaxation rate but concomitantly loses intensity due to depopulation of the triplet state. For a finely ground powder of copper acetate, this is seen in Figure 11.6 where the low-temperature (48 K) spectrum is overwhelmed by the $S = 1/2$ signal of a trace amount of monomeric Cu(II). At liquid helium temperature the triplet EPR has undetectably low intensity because essentially all molecules are in the $S = 0$ state. Spectral analysis based on the "system" Hamiltonian in Equation 11.14 for $S = 1$ affords the zero-field interaction parameters $D = 0.33$ cm^{-1}, $E = 0.011$ cm^{-1}. Interestingly, the splitting in zero field $\Delta = D \pm E$ (cf. Equation 8.41) is approximately equal to the X-band microwave quantum $h\nu \approx 0.3$ cm^{-1}. In other words, the "forbidden," "$|\Delta m_s| = 2$," "half-field" transition has moved into zero field where the mixing of the $|+1\rangle$ and $|-1\rangle$ levels is maximal, and so the intensity of the transition is also maximal.

Many multiple copper containing proteins (e.g., laccase, ascorbate oxidase, hemocyanin, tyrosinase) contain so-called "type III" copper centers, which is a historical name (cf. Section 5.8 for type I and type II copper) for strongly exchange-coupled Cu(II) dimers. In sharp contrast to the ease with which $S = 1$ spectra from copper acetate are obtained, half a century of EPR studies on biological type III copper has not produced a single triplet spectrum. Why all type III centers have thus far remained "EPR silent" is not understood.

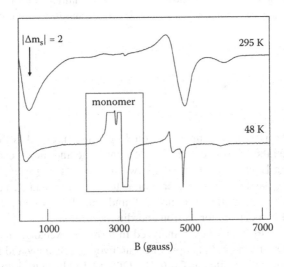

FIGURE 11.6 EPR of the copper dimer in pure copper acetate powder. Strong exchange coupling gives an $S = 0$ ground state and an $S = 1$ excited state at $2J \approx 300$ cm^{-1}. At $T = 48$ K the triplet is hardly populated, and the spectrum is dominated by a trace of monomeric copper.

11.4   SPIN LADDERS

Nature has developed a very rich cluster biochemistry because arranging transition ions together in a single prosthetic group affords a quite significantly broadened tunability of reduction potentials (e.g., ferredoxins) and a remarkably increased catalytic reactivity (e.g., activation of small molecules like H_2, N_2, CO, O_2, N_2O). There is also a bonus for the bioEPR spectroscopist as clustering also brings along rich biomolecular magnetism especially in terms of possible spin states. Thus far in this chapter we have only considered interaction between two $S = 1/2$ centers, but high-spin ions are common, and so are multinuclear interactions (i.e., between more than two metal ions). This section is about the bookkeeping of spin states in clusters. To this goal we need to introduce one more type of interaction known under the somewhat misleading, indeed cryptic, name of "double exchange."

Iron is a common element in biological clusters, in which it is very predominantly high-spin ferric (Fe^{III} with $S = 5/2$) or high-spin ferrous (Fe^{II} with $S = 2$). In iron–sulfur proteins the dinuclear cluster [2Fe-2S] occurs in two possible oxidation states with overall cluster valence 2+ (two ferric ions) or 1+ (a ferric and a ferrous ion), and their ground-state spin is $S = 0$ or $S = 1/2$, respectively, that is, low spin. Apparently, the spins of the individual iron ions are coupled in an antiparallel way into a system spin of minimal value. Dinuclear iron–oxo clusters (two Fe bridged by at least one oxygen either from an amino acid like aspartate and/or as an external ligand such as OH^-) occur in three possible oxidation states: the two irons can be (III, III), (III, II), or (II, II). Just like the iron–sulfur clusters, usually the all-ferric cluster is $S = 0$ and the mixed-valence cluster is $S = 1/2$, however, the all-ferrous cluster has an integer-spin ground state. Several dinuclear iron–oxo model clusters have been synthesized whose mixed-valence ground state has $S = 9/2$, that is, the ferric spin and the ferrous spin are coupled in a parallel manner (Ding et al. 1990). What determines the way in which the spins of the ions couple into a system spin?

There is something peculiar about mixed-valence clusters. Suppose we start from a fully oxidized dinuclear cluster in which we have labeled the irons Fe_A and Fe_B, and we add a single electron (i.e., one equivalent of a reductant with a reduction potential, E'^0 well below that of the $Fe^{III}Fe^{III}/Fe^{III}Fe^{II}$ couple). Which one of the two iron ions will the electron go to? If the cluster happens to be a symmetric model compound, then the ambiguity is maximal: the two possibilities $Fe_A^{III}Fe_B^{II}$ and $Fe_A^{II}Fe_B^{III}$ are fully equivalent, and we should expect some form of resonance stabilization. Alternatively, if the cluster is part of an asymmetric structure, such as a protein, then the difference in relative stability of the two configurations could well be sufficiently large for the occurrence of one to be fully dominant. How does this difference work out in the magnetism and, therefore, in the EPR spectroscopy?

Zener appears to have been the first to consider this problem to some depth in his theoretical work on ferromagnetic crystals of the type $La_{1-x}Ca_xMnO_3$ (Zener 1951). For $x = 0$ one has $La^{III}Mn^{III}O_3$ but for $x > 0$ some of the Mn will be 4+, and so we have the structure $La_{1-x}^{III}Ca_x^{II}Mn_{1-x}^{III}Mn_x^{IV}O_3$ in which some Mn-Mn pairs will be mixed valence, that is, $Mn_A^{III}Mn_B^{IV}$ or $Mn_A^{IV}Mn_B^{III}$. Mn^{III} is $3d^4$ ($S = 2$) and Mn^{IV} is $3d^3$ ($S = 3/2$), and Zener proposed that the excess electron (also called itinerant electron or Zener electron) on Mn^{III} can "travel" to the Mn^{IV} via a doubly-occupied p-orbital of a bridging oxygen. Starting from the two energetically degenerate states, $Mn_A^{III}Mn_B^{IV}$ or $Mn_A^{IV}Mn_B^{III}$, the traveling of the electron would create a new ground state of lower energy in which Mn_A and Mn_B are no longer discernible. For this event Zener cast the name "double exchange"; the key point of his proposal is that the energy lowering only occurs when the original spins of the two partners are parallel. In later work this requirement was generalized to "not completely antiparallel" (Anderson and Hasegawa 1955; De Gennes 1960); see below. Then, Papaefthymiou et al. adopted the concept of double-exchange interaction for truly isolated clusters (namely: in proteins) in their analysis of the magnetism of a $[3Fe-4S]^0$ cluster in a reduced ferredoxin (Papaefthymiou et al. 1987), and the idea was subsequently rapidly adopted by the bioinorganic community at large. In the meantime discussions on the relevance of double exchange continued to develop in the physics literature, with a boost by the more recent interest in the magnetic properties of nanoscale molecular magnets. In fact, after nearly six decades the theme appears to have retained its controversial character until this day (Bastardis et al. 2007).

Against this volatile background the biomolecular spectroscopist should perhaps take a humble position at the side line, however, completely ignoring the subject is not well possible where double exchange has become part of the bioEPR language to rationalize spin multiplicity of metal clusters in proteins and model compounds. We limit the discussion to a basic (especially, isotropic) outline of the matter. Superexchange and double exchange can be viewed as opposing interactions, in the sense that the first leads to a ground state with antiparallel coupling of local spins and the second to parallel coupling. If the first term dominates in a dinuclear cluster, then the result is a minimal system spin for the ground state; domination of the second term results in a maximal system spin; in case of comparable strength of the two terms, the system spin can take any value, in particular intermediate ones.

Schematically, the double exchange interaction can be written in the form of a spin Hamiltonian operator (working on the system spin S_{AB}—for short, S) as

$$H_S = BVT \tag{11.20}$$

in which T is a "transfer function," for example, converting a dinuclear state $|\alpha\rangle = M_A^{x+1}M_B^x$ into the state $|\beta\rangle = M_A^x M_B^{x+1}$ (M^x stands for a metal ion of valence x), and vice versa. And V is the double-exchange operator, which, when working on either system state $|\alpha\rangle$ or $|\beta\rangle$ affords the eigenvalue $S + 1/2$, for example,

$$VT|\beta\rangle = V|\alpha\rangle = (S+1/2)|\alpha\rangle$$
$$VT|\alpha\rangle = V|\beta\rangle = (S+1/2)|\beta\rangle \tag{11.21}$$

And the combination of Heisenberg superexchange plus Zener double exchange results in zero-field energy levels in terms of the system spin S

$$E(S) = J[S(S+1)] \pm B(S+1/2) \tag{11.22}$$

Depending on the relative magnitudes of $|J|$ and $|B|$ one can discern three regimes as illustrated in Figure 11.7 for the mixed-valence pair $Fe^{III}Fe^{II}$. If $|J| \gg |B|$ then double exchange is negligible and we find a classic, regular spin ladder with the smallest system spin associated with the ground level, and the zero-field splittings between the sports increasing with increasing spin. For $|J| \ll |B|$ the double exchange is dominant, and we find the maximal system spin for the ground state as part of an inverted ladder with mirror image at high energies. For the intermediate regime of $|J| \approx |B|$ a complex, "nested" pattern will be found, and the spin of the ground state is a sensitive function of the ratio $|J/B|$ (Hagen 1992).

The ground state of the mixed-valence $[2Fe-2S]^{1+}$ cluster in proteins and model compounds is always $S = 1/2$; apparently, superexchange is dominant. The ground state of some model compounds with cluster core ($Fe^{III}O_2Fe^{II}$) or ($Fe^{III}(OH)_3Fe^{II}$) is $S = 9/2$, suggesting the double exchange to dominate (Ding et al. 1990). These complexes are centrosymmetrical, and the two Fe sites are fully equivalent consistent with the delocalization of the Zener electron. In other model compounds, and particularly in proteins, dinuclear mixed-valence iron–oxo clusters usually have $S = 1/2$ ground states. However, in ribonucleotide reductase the iron–oxo mixed-valence state has been prepared in an $S = 9/2$ ground state from the all-ferric state by gamma irradiation of the protein in the frozen state at 77 K, followed by "annealing" (i.e., removing stresses by allowing limited mobility) by warming up to 165 K (Davydov et al. 1994). Apparently, minor structural rearrangements are sufficient to go from one extreme ($|J| \gg |B|$) to the other ($|J| \ll |B|$), and this implies that we can expect cluster bioEPR spectroscopy to show a rich palette of system spins.

The first signs of this spin richness are seen in the magnetism of the trinuclear iron–sulfur cluster [3Fe-4S], a distorted cube of alternating Fe and S corners from which one Fe has been removed. In the fully oxidized state, $[3Fe-4S]^{1+}$ all three iron

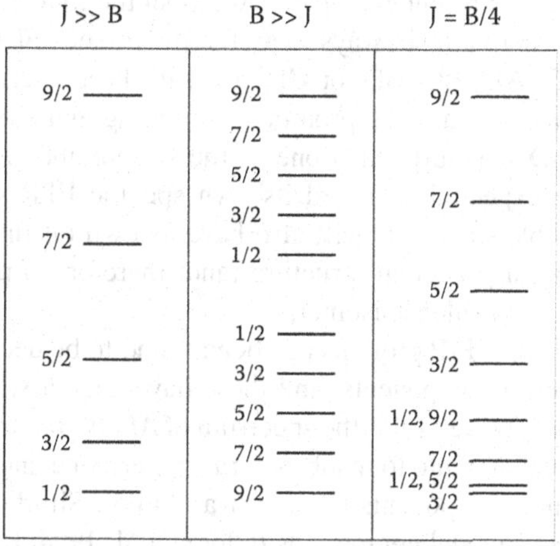

$J \gg B$	$B \gg J$	$J = B/4$
9/2	9/2	9/2
	7/2	
	5/2	
	3/2	7/2
7/2	1/2	
		5/2
	1/2	
5/2	3/2	3/2
	5/2	1/2, 9/2
3/2	7/2	7/2
1/2	9/2	1/2, 5/2 / 3/2

FIGURE 11.7 Spin ladders for the dinuclear Fe^{III}-Fe^{II} cluster. The two metal ions are subject to superexchange (J) and double exchange (B) with $J \gg B$, $J \ll B$, or $J = B/4$. The three ladders are not normalized to the same energy scale.

ions are high-spin ferric. The system spin $S = 1/2$ is thought to result from coupling two Fe $S = 5/2$ spins into an $S = 3$ dimer, and subsequently coupling this structure to the remaining $S = 5/2$ iron into an overall system spin of $S = 1/2$ (Kent et al. 1980, Gayda et al. 1982).

The one-electron reduced cluster, $[3Fe-4S]^0$ has a system spin $S = 2$, which is envisioned to be the result of parallel coupling (i.e., through double exchange) a ferric $S = 5/2$ and a ferrous $S = 2$ ion into a delocalized pair with $S = 9/2$, and subsequently coupling this structure antiparallel to the remaining $S = 5/2$ iron into an overall system spin $S = 2$ (Papaefthymiou et al. 1987).

Frequently, clusters exhibit more than one spin, even when in a single, well defined oxidation state. For example, reduced cubanes in ferredoxins and in complex enzymes can be mixtures of $S = 1/2$ and $S = 3/2$ (Hagen et al. 1985b), the two-electron oxidized [8Fe-7S] cluster in nitrogenase is a mixture of $S = 1/2$ and $S = 7/2$ (Pierik et al. 1993), the oxidized [4Fe-2S-2O] cluster in the hybrid-cluster protein is a mixture of $S = 1/2$ and $S = 9/2$ (Pierik et al. 1992a). Spin counting of these "spin mixtures" usually indicate their integrated intensity to approximately add up to unity, that is, apparently some molecules are low spin and the remainder is high spin. Their ratio does not appear to change with temperature, suggesting that they both represent ground state multiplets. Their ratio also does not appear to change with degree of reduction, suggesting essentially identical chemical structures. The nature of these spin mixtures remains enigmatic.

11.5   VALENCE ISOMERS

The $S = 1/2$ ground state of the oxidized HiPIP cubane, $[4Fe-4S]^{3+}$, is thought to result from the antiparallel coupling of the subspin $S = 4$ or 3 of a ferric dimer and the subspin $S = 9/2$ or 7/2 of a delocalized ferric-ferrous dimer. Whatever the exact coupling scheme, there is generally more than one way to assign substructures to the full structure. In the present example, if we label the four irons of the cubane as $Fe_A Fe_B Fe_C Fe_D$, then there are six ways to assign the mixed-valence dimer to two of these irons: AB, AC, AD, BC, BD, or CD. We call these alternative charge distributions *valence isomers*. Since the protein surrounding and coordinating this cluster has a specific 3-D structure, each one of the six possible mixed-valence dimer assignments can be expected to afford its own specific EPR spectrum. However, each of the six possible structures will also have its own intrinsic stability, and the probability of finding a particular structure (and, therefore, a particular spectrum) will be different for each valence isomer.

The X-band EPR of HiPIP proteins has been found to be deconvolutable in from 1 to 4 different spectral components, and these have been assigned to valence isomers. As an example, Figure 11.8 is the spectrum of *Allochromatium vinosum* HiPIP, which can be deconvoluted into four subspectra, namely, one major component (circa 70%) and three minor components (Priem et al. 2005). Similar "heterogeneity" is found in the spectra of model compounds (Gloux et al. 1994, Le Pape et al. 1997)). It is not known whether the occurrence of valence isomers is a general phenomenon for biological metal clusters. The X-band $S = 1/2$ spectra of the $[4Fe-4S]^{1+}$ cluster in ferredoxins appear to be single-component, however, when the HiPIP of *Rhodopila*

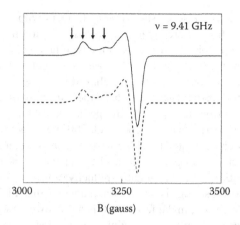

FIGURE 11.8 EPR spectrum of [4Fe-4S]$^{3+}$ valence isomers. The experimental spectrum (solid trace) of *Allochromatium vinosum* HiPIP is simulated as a sum of four slightly different spectra presumably presenting four of the possible six valence isomers. Arrows indicate the g_z's of the four components with relative concentration (from left to right): 8%, 68%, 12%, and 12%.

globiformis is "superreduced" with two reducing equivalents to the 1+ state, the EPR spectrum does exhibit the heterogeneity typical for valence isomerism (Heering et al. 1995).

11.6 SUPERPARAMAGNETISM

In a homogeneous ferromagnet the spins of all the individual magnetic atoms are aligned. Above a certain temperature, called the Curie point, the parallel exchange coupling between the spins is overcome by thermal energy, and the material becomes a paramagnet. In a homogeneous antiferromagnet the spins of all the individual magnetic atoms are aligned antiparallel with neighboring spins. Above a certain temperature, called the Néel point, the antiparallel exchange coupling between the spins is overcome by thermal energy, and the material becomes a paramagnet. If a ferromagnetic or an antiferromagnetic material consists of small crystallites (also called *grains, nanoparticles, Weiss domains*) then the ordering of spins may not be perfectly parallel or antiparallel, as long as the "average correlation" of neighboring spins is still ferromagnetic or antiferromagnetic. In such material the temperature required to overcome coupling between the grains (not between the atoms) is lower than the Curie temperature or Néel temperature. This temperature point is called the *blocking temperature*, and the resulting magnetic state is called *superparamagnetism*. In biology the magnetism of iron storage proteins is interpreted in terms of superparamagnetism of a nonhomogeneous antiferromagnet.

The iron storage protein ferritin is a small 20 kDa α-helical protein that spontaneously assembles into a hollow ball-like homo-24-mer. The outer diameter of the sphere is circa 12 nm and the inner diameter, or core diameter, is circa 8 nm. A smaller version, known as *miniferritin* or Dps protein (Dps = DNA protecting protein during starvation) also exists, forming 12-mers with an inner diameter of circa 4.5 nm. The ball has pores that give access to hydrophilic channels to take up ions, and to hydrophobic channels, presumably to take up, for example, molecular oxygen and/or protons. Each subunit has a binding motif, called the *ferroxidase site*, for a dinuclear iron–oxo cluster. It is believed that in aerobic species Fe^{II} ions enter and bind to the ferroxidase site, where they are pair-wise oxidized to Fe^{III}, whereupon they are further transported into the interior of the ball for deposition as a mineral resembling ferrihydrite (Michel et al. 2007) with approximate stoichiometry $Fe_2O_3 \bullet 0.2H_2O$ or $Fe_{10}O_{14}(OH)_2$. Alternatively, concomitant uptake of iron and oxoanions, in particular phosphate, can lead to deposition of an amorphous core of variable composition. The (presumably reductive) mechanism by which iron is released from its ferritin storage is not known. It appears that all forms of life on this planet, including anaerobes, make ferritins. It is not known how ferritins operate in the absence of oxygen. The maximum number of iron ions stored is circa 3000 per ferritin and circa 500 per miniferritin.

The X-band EPR spectrum of ferritin shows an extremely broad feature at liquid nitrogen temperature and above from the superparamagnetic core (Figure 11.9). Upon lowering the detection temperature, somewhere between 77 K and 4 K, the broad signal completely disappears, and this is interpreted as the transition from superparamagnetism to antiferromagnetism. Estimates for the blocking temperature range from 15 to 38 K (Frankel et al. 1991; Luis et al. 1999; Resnick et al. 2004). Although multiple high-resolution x-ray crystallographic structures have been determined for the protein part of ferritins, the structure(s) of the core remains elusive. Incubation of Dps apo-ferritin crystals with iron and freezing after distinct time intervals, followed by crystallography, has indicated that the nucleation for core formation starts with the formation of a specific cluster consisting of five iron ions (Zeth et al. 2004). This suggests that the final core can perhaps be envisioned as a collection of very small "grains" or antiferromagnetic nanoparticles, but much still remains to be asked for in terms of structural and (EPR) spectroscopic determination.

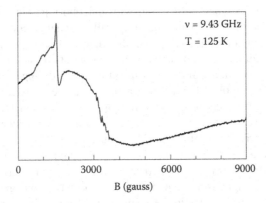

FIGURE 11.9 An extremely broad EPR signal form the superparamagnetic core in ferritin. The spectrum is from *Pyrococcus furiosus* ferritin. The sharp signal at $g = 4.3$ (circa 1570 gauss) is from a trace of contaminating "dirty iron."

12 High Spins Revisited

In the theoretical Chapters 7–9 we dived into the matter of high spins to a considerable depth. We looked at half-integer spins in the weak-field limit in Chapter 7, and we derived the rhombogram for $S = 3/2$ both in the weak-field limit and for the more involved intermediate-field case. Earlier, in Chapter 5 we already scrutinized the $S = 5/2$ rhombogram in the weak-field limit with Fe-SOD (superoxide dismutase) as an example. We considered integer-spin systems and in Chapter 8 we identified an unexpected prominence for the formally forbidden transition within the non-Kramer's doublet with the highest m_S-values. In Chapter 9 we introduced g-strain and A-strain, but we decided to postpone development of a description of D-strain in high-spin systems to the present chapter. Compared to the heavy QM treated in Chapter 7–9, this chapter is a rather easy-going follow-up excursion into the realm of high-spin bioEPR. Here, we are particularly interested in practical aspects of the spectral analysis, and the extension of the theory required in addition to what we already know will be minor.

12.1 RHOMBOGRAMS FOR $S = 7/2$ AND $S = 9/2$

High-spin systems are at least subject to the electronic Zeeman interaction and to the zero-field interaction; see Equation 11.4. We make the following assumptions: (1) only the zero-field S^2 terms are significant for half-integer spins, as was discussed in Section 8.1; (2) the g and D-tensors are colinear; and (3) the real g-values are approximately known, e.g., $g_x \approx g_y \approx g_z \approx g_e$ (see below). Then, in the weak-field limit, $S*S \gg B*S$, the effective g-values of intra-doublet transitions can be presented in rhombograms of g^{eff} versus $\eta = E/D$. The rhombogram for $S = 3/2$ was given in Chapter 7, Figure 7.1, and the one for $S = 5/2$ is in Chapter 5, Figure 5.10. To this collection we now add the rhombograms for $S = 7/2$ in Figure 12.1 and for $S = 9/2$ in Figure 12.2. The plots have been calculated with the procedure outlined in Chapter 8 for $S = 3/2$. The required energy matrices for $S = 7/2$ (Hagen et al. 1987) and $S = 9/2$ (Pierik and Hagen 1991), based on a spin Hamiltonian with a D- and a g-tensor, are given here for the record

$\langle +7/2\|$	$7D+7G_z$	0	$\sqrt{7}G_-$	0	$\sqrt{21}E$	0	0	0
$\langle -7/2\|$	0	$7D-7G_z$	0	$\sqrt{7}G_+$	0	$\sqrt{21}E$	0	0
$\langle +5/2\|$	$\sqrt{7}G_+$	0	$D+5G_z$	0	$2\sqrt{3}G_-$	0	$\sqrt{45}E$	0
$\langle -5/2\|$	0	$\sqrt{7}G_-$	0	$D-5G_z$	0	$2\sqrt{3}G_+$	0	$\sqrt{45}E$
$\langle +3/2\|$	$\sqrt{21}E$	0	$2\sqrt{3}G_+$	0	$-3D+3G_z$	0	$\sqrt{15}G_-$	$2\sqrt{15}E$
$\langle -3/2\|$	0	$\sqrt{21}E$	0	$2\sqrt{3}G_-$	0	$-3D-G_z$	$2\sqrt{15}E$	$\sqrt{15}G_+$
$\langle +1/2\|$	0	0	$\sqrt{45}E$	0	$\sqrt{15}G_+$	$2\sqrt{15}E$	$-5D+G_z$	$4G_-$
$\langle -1/2\|$	0	0	0	$\sqrt{45}E$	$2\sqrt{15}E$	$\sqrt{15}G_-$	$4G_+$	$-5D-G_z$

$$(12.1)$$

$$\begin{array}{r}\langle +9/2| \\ \langle -9/2| \\ \langle +7/2| \\ \langle -7/2| \\ \langle +5/2| \\ \langle -5/2| \\ \langle +3/2| \\ \langle -3/2| \\ \langle +1/2| \\ \langle -1/2|\end{array}\begin{bmatrix}12D+9G_z & 0 & 3G_- & 0 & 6E & 0 & 0 & 0 & 0 & 0 \\ 0 & 12D-9G_z & 0 & 3G_+ & 0 & 6E & 0 & 0 & 0 & 0 \\ 3G_+ & 0 & 4D+7G_z & 0 & 4G_- & 0 & 2\sqrt{21}E & 0 & 0 & 0 \\ 0 & 3G_- & 0 & 4D-7G_z & 0 & 4G_+ & 0 & 2\sqrt{21}E & 0 & 0 \\ 6E & 0 & 4G_+ & 0 & -2D+5G_z & 0 & \sqrt{21}G_- & 0 & 3\sqrt{14}E & 0 \\ 0 & 6E & 0 & 4G_- & 0 & -2D-5G_z & 0 & \sqrt{21}G_+ & 0 & 3\sqrt{14}E \\ 0 & 0 & 2\sqrt{21}E & 0 & \sqrt{21}G_+ & 0 & -6D+3G_z & 0 & \sqrt{24}G_- & 5\sqrt{6}E \\ 0 & 0 & 0 & 2\sqrt{21}E & 0 & \sqrt{21}G & 0 & -6D-3G_z & 5\sqrt{6}E & \sqrt{24}G_+ \\ 0 & 0 & 0 & 0 & 3\sqrt{14}E & 0 & \sqrt{24}G_+ & 5\sqrt{6}E & -8D+G_z & 5X_- \\ 0 & 0 & 0 & 0 & 0 & 3\sqrt{14}E & 5\sqrt{6}E & \sqrt{24}G & 5X_+ & -8D-G_z\end{bmatrix}$$

(12.2)

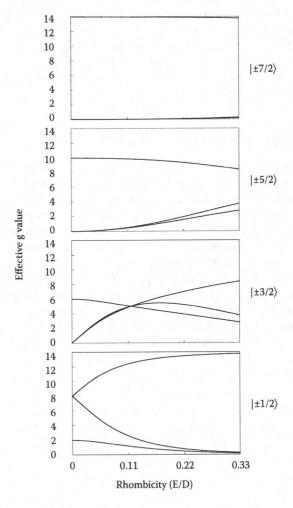

FIGURE 12.1 Rhombogram for $S = 7/2$. A plot of the effective g-values of the four intra-doublet transitions as a function of the rhombicity $\eta = E/D$ assuming $g_{real} = 2.00$ and $S*S \gg S*B$.

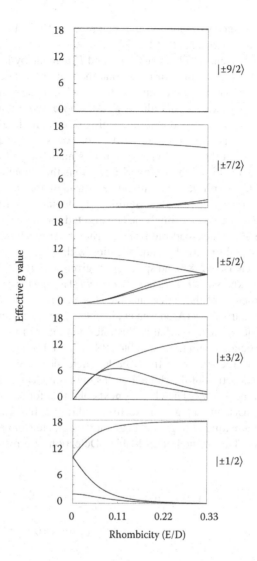

FIGURE 12.2 Rhombogram for $S = 9/2$.

Why do we not extend our inventory beyond $S = 9/2$? A practical answer is that half-integer spin systems with $S > 9/2$ appear to be rare in biology. In fact, at the time of writing I know of no example, but note that in the field of molecular magnets the number of identified $S > 9/2$ systems is rapidly increasing. We can also argue why we do not expect to find very high spins in biology. For mononuclear systems (one metal ion) the maximal spin is $S = 5/2$ for d^5 systems (Fe^{III}, Mn^{II}). For pairs of d-ions the maximal half-integer spin $S = 9/2$ is found for the mixed-valence configuration (M^{III}-M^{II}). Such a pair should be in a fairly symmetrical coordination surrounding in order for double exchange to be dominant and thus to afford such a high spin. Apparently, the intrinsic low symmetry of biomolecules combined with the limited size of metal clusters in biomolecules ensures that $S \le 9/2$.

The assumption of tensor colinearity for g and D is made by lack of proof to the contrary, although it is probably fair to say that the matter has not been explored to any significant depth in bioEPR thus far. On the other hand, the richness of high-spin bioEPR is particularly associated with the d^5 ion whose real g-tensor is essentially isotropic and nearly equal to g_e due to quenching of orbital angular momentum in half-filled shell systems. An isotropic tensor is of course colinear with any other tensor. Furthermore, in the mixed-valence clusters made of Fe^{III} ($S = 5/2$) and Fe^{II} ($S = 2$) ions or of Mn^{III} ($S = 2$) and Mn^{II} ($S = 5/2$) ions the deviations from g_e in the g-values of the integer-spin ions are attenuated through spin couplings to the $S = 5/2$ ions. And if the real g-values do significantly deviate from g_e (e.g., Co^{II}, $S = 3/2$) then the rhombograms can simply be constructed for these specific g_{real}-values. Alternatively, the g^{eff}-values read-out from the rhombograms based on $g_{real} = g_e$ can simply be extrapolated linearly by multiplication with g_{real}/g_e.

An example of an $S = 9/2$ X-band spectrum is given in Figure 12.3. The spectrum is from the enzyme known as *hybrid cluster protein* (HCP) whose trivial name refers to the unusual structure of the active-site 4Fe cluster, which contains both S and O bridges between the iron ions (Arendsen et al. 1998). The spectrum in Figure 12.3 is a "typical" example in the sense that it illustrates a number of the complexities that are frequently associated with high-spin bioEPR. The first thing to notice is that the very first "peak" in the spectrum from the low-field side has an effective g-value greater than 14. Inspection of the rhombograms makes it clear that this observation defines the spin to be $S \ge 9/2$. The obvious next condition for us to check is whether the spectrum is consistent with a spin exactly equal to $9/2$. In the rhombogram for $S = 9/2$ we find that a feature with $g^{eff} > 14$ can only originate from a transition within either the $|\pm 1/2\rangle$ doublet or the $|\pm 9/2\rangle$ doublet. Detection of a resonance within the

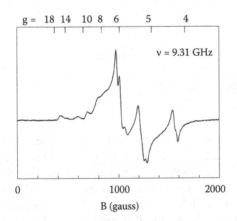

FIGURE 12.3 $S = 9/2$ EPR of *Desulfovibrio desulfuricans* hybrid cluster protein. Note the weak low-field peak with $g > 14$ and the relatively strong intensities in the $g \approx 5.4$ and $g \approx 6.4$ regions, which are all characteristics of the $S = 9/2$ system.

latter doublet is unlikely because the anisotropy in g^{eff} is seen to be quite enormous over the whole range of possible rhombicities (cf. Figure 12.2). For a transition within the $|\pm1/2\rangle$ doublet the amplitude of the $g^{eff} \approx 16$ line is rather weak, indicating that the $S = 9/2$ multiplet might be inverted (i.e., $D < 0$ and the $|\pm1/2\rangle$ doublet is highest). This is indeed borne out by temperature-dependent measurements: at 4 K the line has disappeared (Pierik et al. 1992a). The value of $g^{eff} \approx 16$ defines a rhombicity of $\eta \approx 0.08$, and this value in turn predicts a number of resonances in the g^{eff}-range of approximately 5–7, which should have significant intensity due to the relatively limited g-anisotropy. This is indeed what we find qualitatively in Figure 12.2, however, the set of observed resonances in this $g^{eff} \approx 5$–7 range is not consistent with a single unique value of the rhombicity, which indicates that either the sample is not completely homogeneous (resulting in a multiplicity of rhombicities), or the real g-tensor is anisotropic (and therefore all real g-values differ from g_e), or both conditions are true. Finding multiple rhombicities in the high-spin EPR of otherwise apparently homogeneous metalloproteins is not uncommon. The spectrum in Figure 12.3 also has a feature at $g^{eff} \approx 4.3$ that is probably from a minor "dirty iron" $S = 5/2$ contaminant; it is not predicted by the rhombogram for $\eta \approx 0.08$. All in all, the spectrum is broadly consistent with $S = 9/2$ if we allow for multiple values for the rhombicity of the order of $\eta \approx 0.08$.

The example, above, illustrates several points of general relevance. In Chapter 5 we have previously come across the mathematical coincidence of an isotropic $g^{eff} = 4.29$ for an $S = 5/2$ system of maximal rhombicity $\eta = 1/3$ (Figure 5.13). We now see that similar "mathematical coincidences" can occur for higher spins, as listed in Table 12.1. All weak-field half-integer spin systems with $S = 1/2 + 2n$ ($n = 0, 1, 2$, etc.) have an isotropic transition within the middle doublet when the rhombicity is maximal. This is a reflection of a mirror symmetry property of rhombograms. At maximal rhombicity the effective g-values of the lowest doublet are identical to those of the highest doublet; equally, the effective g-values of the one-to-lowest doublet are identical to those of the one-to-highest doublet, and so on. Systems with $S = 1/2 + 2n$ have an odd number of doublets, and, therefore, an odd number of subrhombograms. For these systems the middle subrhombogram is mirrored into itself around the $\eta = 1/3$ axis, and the three effective g-values must coincide at the mirror axis. The practical importance of these isotropies should be obvious from the ubiquitous occurrence of the $g = 4.3$ "dirty iron" signal in all EPR spectra of biological samples:

TABLE 12.1
Isotropic effective g-values in high-spin systems

Spin	Isotropic g^{eff}	Doublet	Rhombicity η
3/2	none	—	—
5/2	4.29	Middle	Maximal
7/2	5.01	Second	0.117
9/2	5.35	Second	0.055
	6.36	Middle	Maximal

if a weak-field half-integer spin system occurs with a rhombicity close to its isotropy value listed in Table 12.1, then its spectrum is likely to be readily detectable as it is dominated by an isotropic resonance with g-value given in the table. Moreover, even if only a small fraction of all biomolecules in a sample have $\eta = 1/3$, then the isotropic line will still be the main (i.e., highest-amplitude) feature of the spectrum. This general phenomenon is analyzed in quantitative terms in the next section.

12.2 D-STRAIN MODELED AS A RHOMBICITY DISTRIBUTION

In Chapter 9 we extensively discussed the consequences of protein conformational distributions for the shape of powder EPR patterns under the label "g-strain." At the end of that chapter we also briefly looked at how g-strain indirectly manifests itself through changes in hyperfine interaction patterns under the name of A-strain, but we decided to postpone considering possible strain effects through the zero-field interaction (D-strain). Now that we have completed our inspection of rhombograms and have come to realize the general importance of coincidental isotropy, we are ready to develop a practical description of D-strained spectra. Rigorous statistical and/or molecular models for D-strain in metalloproteins do not exist at this time. We will develop a very simple, phenomenological model, whose value will be underpinned by its ability to semiquantitatively generate powder patterns that fit experimental data (Hagen 2007).

The zero-field spin Hamiltonian parameters, D and E, are assumed to be distributed according to a normal distribution with standard deviations σ_D and σ_E, which we will express as a percentage of the average values $\langle D \rangle$ and $\langle E \rangle$. g-Strain itself is not expected to be of significance, because the shape of high-spin spectra in the weak-field limit is dominated by the zero-field interaction.

If the distributions in D and E would be fully positively correlated, then no broadening of the EPR spectrum whatsoever would ensue, because their ratio (i.e., the rhombicity $\eta = E/D$) would remain constant, and so the effective g-values would not be distributed. On the other hand, a maximal broadening effect would occur in case of full negative correlation: $\sigma_D = -\sigma_E$. This definition reduces the description of D-strain to the fitting of a single parameter.

In practice, the average of D is set to a high dummy value, $D \gg 0.3$ cm^{-1}, for example $|D| \equiv 10$ cm^{-1}. The average of E is then defined through the rhombicity, $\langle E \rangle = \eta \langle D \rangle$, and the sign of E is taken to be opposite to that of D. The fully negatively correlated normal distributions in D and E lead to an asymmetric distribution in the rhombicity $|\eta|$, defined by a single parameter $\sigma = |\sigma_D| = |\sigma_E|$. The rhombicity itself is estimated from the experimental spectrum by moving a vertical ruler over the rhombogram in search of a match to (some of) the effective g-values. The procedure to simulate powder patterns then simply involves fitting the distribution parameter σ and fine-tuning the average value of η. To obtain the strained EPR powder spectrum, for each set of (D,E) values, the energy matrix for the spin at hand is diagonalized for each of the three canonical orientations (B along a molecular axis), and the effective $S = 1/2$ resonance condition for each doublet is solved. All the resulting spectra for the different values of η are summed with proper weighting according to the rhombicity distribution to obtain the D-strained overall spectrum.

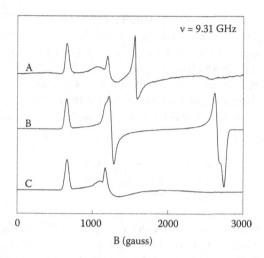

FIGURE 12.4 $S = 7/2$ EPR of the [8Fe-7S] P-cluster in *Azotobacter vinelandii* nitrogenase. The experimental spectrum (trace A) has been simulated in the absence (trace B) and the presence (trace C) of D-strain modeled as a correlated distribution in the zero-field parameters D and E.

To illustrate the power of this simple ad hoc model, consider the $S = 7/2$ EPR of the [8Fe-7S] P-cluster (the name refers to the fact that the cluster is protein-bound and not removable without destruction) of the N_2-activating enzyme nitrogenase. This oxygen-sensitive enzyme is purified in the presence of reductant dithionite. After removal of the dithionite and oxidation of the enzyme by two equivalents, the P-cluster has $S = 7/2$ (Hagen et al. 1987; Pierik et al. 1993). The X-band spectrum is in Figure 12.4. Reading out the field positions of the spectral features, converting them to g^{eff}-values (10.4, 5.8, 5.5), and fitting these to a vertical line in the $S = 7/2$ rhombogram in Figure 12.1 gives a rhombicity $\eta = 0.044$. When we now measure the experimental linewidth of the line at lowest field, and then generate subspectra with constant linewidth of all intra-doublet transitions, and add these up to construct a simulation of the experimental spectrum, we find a rather poor fit in terms of positions as well as intensities especially towards higher field values where the simulation predicts a sharp, intense feature that is not found experimentally. On the contrary, using an 8% distribution width for (D,E) in the single-parameter strain model affords a very reasonable fit to the experimental spectrum (Figure 12.4).

12.3 POPULATION OF HALF-INTEGER SPIN MULTIPLETS

The experimental spectrum in Figure 12.4 was recorded for a sample temperature of circa 20 K. Upon lowering of the temperature, and concomitantly reducing the microwave power to avoid saturation, one finds that the normalized intensity of the spectral features diminishes. At 10 K the lines with $g^{eff} = 10.5$ and 5.5 are gone and only the $g^{eff} = 5.8$ peak remains; at 4.2 K also this latter feature has disappeared: the doublets that give rise to these transitions ($|\pm 1/2\rangle$ and $|\pm 3/2\rangle$, respectively) have been depopulated. They are excited states within the $S = 7/2$ manifold because the axial zero-field splitting parameter has a negative value: $D \approx -3.7$ cm^{-1} (Hagen et al. 1987). Figure 12.5 gives a plot of the fractional population of the $|\pm1/2\rangle$ doublet (the highest one), together with the normalized amplitude data of the $g^{\mathrm{eff}} = 10.4$ peak. Actually, the value of D was deduced from this fit to the experimental $g^{\mathrm{eff}} = 10.4$ amplitude data. Strictly speaking, the graphs are only correct in zero field: they are calculated with the Boltzmann distribution (Equation 4.1) and the axial splittings in zero field (Figure 5.12), i.e.:

$$\frac{n_i}{n_0} = \frac{\exp(-\Delta E_i / kT)}{\displaystyle\sum_{i=0,1,2,3} \exp(-\Delta E_i / kT)} \qquad (12.3)$$

in which n_0 refers to the ground doublet (here, the $|\pm7/2\rangle$ doublet), n_1 to the first excited doublet ($|\pm5/2\rangle$), and so on. From Figure 5.9 (alternatively, from the diagonal in Equation 12.1) one can see that the energies for an inverted $S = 7/2$ multiplet are $\Delta E_0 = 0$, $\Delta E_1 = 6D$, $\Delta E_2 = 6D + 4D$, $\Delta E_3 = 6D + 4D + 2D$. The intensity values for the $g^{\mathrm{eff}} = 10.4$ peak will slightly deviate from the calculated values because the actual energy levels are slightly shifted due to the small E-term and to the Zeeman interaction, but these deviations are usually well within the relatively large experimental error (typically 20–30%) of experiments of the type presented in

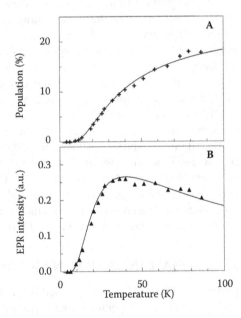

FIGURE 12.5 Temperature-dependent population of the $|\pm1/2\rangle$ doublet of an $S = 7/2$ system. The area under the low-field peak in Figure 12.4 has been fit to the Boltzmann distribution for $S = 7/2$ with $D = -3.7$ cm^{-1}.

Figure 12.5. If desired, these small effects can easily be exactly included by letting the simulation program output all the relative energy levels for the experimentally used frequency and for the canonical orientation that corresponds to $g^{eff} = 10.4$, and then using these values in Equation 12.3. The outcome of Equation 12.3 for a given experimental temperature is required for spin counting (determination of the concentration of the $S = 7/2$ system) using the single-peak integration procedure explained in Section 6.2.

Equation 12.3 is readily rewritten for any other half-integer spin. For integer spins we have one singulet (the $m_S = 0$ level) and S doublets, but we do not bother writing down the modified equation, because spin counting for integer-spin systems in the weak-field limit is near-to-impossible (Hagen 2006).

12.4   INTERMEDIATE-FIELD CASE FOR $S = 5/2$

The intermediate-field case (i.e., $S*S \approx S*B$) is rather common for inorganic systems in X-band, e.g., transition ions doped in diamagnetic Al_2O_3, but the literature holds few data on biological systems. This may be the case because such biosystems happen to be rare or because not many have been found yet due to intrinsic difficulties in the detection of their EPR. The latter possibility perspires from a limited number of studies of five-coordinated Mn^{II} systems with $S = 5/2$ and D-values of the order of the X-band microwave quantum. The key problem here is low spectral intensity (or its mirror image: limited concentration sensitivity of the spectrometer), which is readily illustrated on the example of manganese phosphoglucose isomerase in Figure 12.6. The free enzyme exhibits the well-known six-line pattern of Mn^{II} with a small zero-field splitting ($D \approx 0.01$ cm^{-1}; $S*S \ll S*B$). Upon binding of the substrate fructose-6-phosphate the Mn coordination changes drastically, and the zero-field splitting increases drastically ($D \approx 0.18$ cm^{-1}; $S*S \approx S*B$). With the used substrate concentration, the binding is not quantitative and circa 20% of the enzyme remains free. The signal amplitude of this 20% is found to be circa 40 times greater than that of the 80% bound form. In other words, for equal concentration of the forms the $D = 0.01$ cm^{-1} spectrum has two orders of magnitude greater amplitude than the $D = 0.18$ cm^{-1} form (Berrisford et al. 2006).

In a study on another manganese enzyme, glutathione transferase, the Hoffmann group has proposed Q-band dispersion EPR at the unusually low temperature of 2 K as the optimal approach to collect data from Mn^{II} centers with $D \approx h\nu$ (Smoukov et al. 2002). This proposal would be practically limited by the fact that Q-band spectrometers running at 2 K can be counted on the fingers of one finger; however, dispersion spectra are readily obtained in Q-band also at helium-flow temperatures (i.e., $T > 4.2$ K).

A potentially interesting aspect of the X-band (in contrast to Q-band) is the ready availability of parallel-mode resonators: these types of spectra ($S*S \approx S*B$) have parallel-mode spectra of intensity comparable to the normal-mode spectra (cf. Figure 12.7), and so parallel-mode EPR is an easy way to obtain an independent data set for spectral analysis. This interesting aspect of the intermediate-field case remains to be explored and developed.

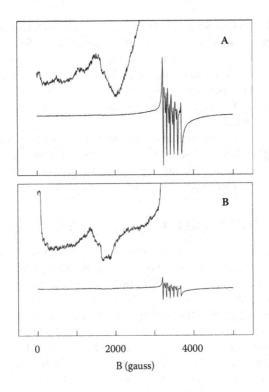

FIGURE 12.6 EPR of an $S = 5/2$ system in the $S*S \approx S*B$ regime. The spectra are from high-spin Mn[II] in *Pyrococcus furiosus* phosphoglucose isomerase in the absence (A) and presence (B) of the substrate fructose-6-phosphate. The substrate-bound, presumably pentacoordinate Mn center gives rise to the broad feature of low intensity. The substrate-free hexacoordinated form affords the strong sextet signal. In box B the two signals differ in amplitude by two orders-of-magnitude; however, their spin count is approximately equal.

12.5   ANALYTICAL LINESHAPES FOR INTEGER SPINS

In Chapter 8 we discussed the particular relevance of *intra*-doublet zero-field splittings in non-Kramer's systems in relation to the effect of higher-order spin operators. EPR of integer-spin systems, especially those of biological origin, typically exhibit X-band spectra only by virtue of a splitting Δ within the highest non-Kramer's doublet being limited by $0 < \Delta < h\nu$. In passing, we have also noticed that the resulting spectral features are usually rather asymmetric (Figures 5.14 and 8.1). These two observations were originally made and combined in the seminal work of Bleaney and coworkers on non-Kramer's lanthanide ions (Bleaney and Scovil 1952, Bleaney et al. 1954, Baker and Bleaney 1958) to develop an effective analytical description of the resonance condition and the lineshape. They described

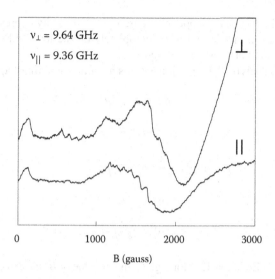

$\nu_\perp = 9.64$ GHz

$\nu_{||} = 9.36$ GHz

$\perp$

$||$

0   1000   2000   3000

B (gauss)

FIGURE 12.7 Dual-mode EPR of mononuclear manganese in phosphoglucose isomerase. The figure shows that half-integer high-spin systems in the $S*S \approx S*B$ regime can have significant intensity in parallel-mode EPR.

the transition in the Δ-split $|\pm m_S\rangle$ doublet of an $S = 2$ system as an *effective* $S = 1/2$ system with resonance condition

$$hv = \sqrt{\Delta^2 + \left(g^{eff}\beta B_0\right)^2} \qquad (12.4)$$

which follows directly from the energy matrix written for an isolated doublet with zero-field splitting Δ:

$$\begin{array}{c} \langle +1/2| \\ \langle -1/2| \end{array} \left\| \begin{array}{cc} (\Delta/2)+G & 0 \\ 0 & -(\Delta/2)-G \end{array} \right. \qquad (12.5)$$

with $G = g^{eff}\beta B_0 / 2$ and $g^{eff} = g_{||}^{eff}\cos\theta$, in which θ is the familiar polar angle defined in Figure 6.4, and for the $|\pm 2\rangle$ doublet $g_{||}^{eff} = 4g_{||} \approx 8$ and $g_\perp^{eff} = 0$ (which makes the off-diagonal elements in Equation 12.5 equal to zero). The transition probability is (Baker and Bleaney 1958)

$$I \propto B_{1_{||}}^2 \left(\frac{g^{eff}\beta\Delta}{2hv} \right)^2 F\left(\Delta^2 / \Delta_0^2\right) d\left(\Delta^2 / \Delta_0^2\right) \qquad (12.6)$$

in which the prefactor B_1 is the microwave magnetic field *parallel* to the static field, and F is determined by a symmetrical gaussian distribution in Δ around the average value $\langle\Delta\rangle = 0$. The fact that Δ appears squared in the intensity expression causes the lineshape to be asymmetric and to be completely on the low-field side of g^{eff} with *zero* intensity at the field corresponding to g^{eff}.

To deduce an analytical expression for this asymmetrical lineshape we define

$$B_g \equiv h\nu / g^{\text{eff}}\beta$$
$$b \equiv B_g - B$$
(12.7)

which gives

$$\left(g^{\text{eff}}\beta B\right)^2 = \left(g^{\text{eff}}\beta B_g - g^{\text{eff}}\beta b\right)^2$$
$$= \left(h\nu - g^{\text{eff}}\beta b\right)^2$$
$$= \left(h\nu\right)^2 - 2g^{\text{eff}}\beta b h\nu + \left(g^{\text{eff}}\beta b\right)^2$$
(12.8)

and substitution in Equation 12.4 for a generalized value of B (instead of B_0) then gives

$$\Delta^2 = 2g^{\text{eff}}\beta b h\nu - \left(g^{\text{eff}}\beta b\right)^2$$
(12.9)

In their subsequent analysis Baker and Bleaney (ibidem) decided to ignore the last term on the assumption that $g^{\text{eff}}\beta b \ll h\nu$. Although this is a reasonable approximation for lanthanide and actinide integer-spin ions doped in single crystals, it is not usually an acceptable assumption for the broad-line spectra from metalloproteins. Furthermore, the assumption of a Δ-distribution around zero (i.e., $D \neq 0$ but all other zero-field interaction parameters are zero) is equally untenable for biomolecules. Therefore, we go for a later extension of the theory, based on a full Equation 12.9 and on $\langle\Delta\rangle \neq 0$, for application to metalloproteins (Hagen 1982b).

With the additional definitions

$$b_0 \equiv B_g - B_0$$
(12.10)
$$p \equiv b(2 - b / B_g)$$

and the corollaries

$$\Delta_0^2 = 2g^{\text{eff}}\beta b_0 h\nu - \left(g^{\text{eff}}\beta b_0\right)^2$$
$$p_0 = b_0(2 - b_0 / B_g)$$
(12.11)

the lineshape for $\langle\Delta\rangle = 0$ becomes

$$I(B) = \int_0^{B_g} \frac{(g^{\text{eff}})^2 p}{\Delta_0^2} \exp\left(-\frac{p}{p_0}\right) dB$$
(12.12)

in which B_g is the axial resonance field where the signal intensity is zero, and b_0 is the shift in field units from B_g down to the field B_0 where the asymmetric EPR absorption line has maximal intensity.

If the intradoublet splitting is distributed around a nonzero value Δ_r, i.e., if a rhombic E-term and/or higher order cubic terms (cf. Section 8.1) are nonzero, then we have

$$\Delta_r^2 = 2g^{eff}\beta b_r h\nu - \left(g^{eff}\beta b_r\right)^2$$

$$p_r = b_r(2 - b_r / B_g) \tag{12.13}$$

$$q_r = \left(\sqrt{p} - \sqrt{p_r}\right)^2$$

and the exponential in the lineshape function changes:

$$I(B) = \int_0^{B_g} \frac{(g^{eff})^2 p}{\Delta_0^2} \exp\left(-\frac{q_r}{p_0}\right) dB \tag{12.14}$$

The signal is maximal for $\theta = 0$, and rotating B (and B_1) away from the molecular z-axis causes a rapid divergence of linewidth and intensity. Thus, the powder spectrum is obtained as the integral

$$S(B) = \int_0^1 d\cos\theta \int_0^{B_g} I(\theta, B) dB \tag{12.15}$$

in which we implicitly use $g^{eff} = g_{\parallel}^{eff}\cos\theta$ and its implied equalities $B_g = B_{g_\parallel}\cos\theta$, $b_0 = b_{0_\parallel}\cos\theta$, and $b_r = b_{r_\parallel}\cos\theta$. The practical value of Equation 12.15 is in the fact that it is an analytical expression, which means that it can be implemented in a spectral simulator to generate accurate powder spectra on standard PC equipment in a split second, which would be a hard act to follow for a simulator based on energy-matrix diagonalization.

Before we ease up by looking at an example, there is one more formal statement to be made, and this is a really important one. While in parallel-mode EPR ($B_1 \parallel B$) the intensity of intradoublet transitions for the canonical orientation $\theta = 0$ is maximal, in regular, perpendicular-mode EPR ($B_1 \perp B$) it is zero. Only when we turn away from the molecular z-axis, the regular spectrum gets finite intensity. By consequence, the regular spectrum is generally broader and of lower intensity than the parallel-mode spectrum. The intensity ratio is

$$I_\perp / I_\parallel = \frac{1}{2}\tan^2\theta \tag{12.16}$$

Inclusion of this factor in Equation 12.15 affords the normal-mode powder spectrum.

Figure 12.8 shows an illustration of the above theory on a very simple model compound: the Fe^{II} ion in (frozen) water. The high-spin ferrous ion is $S = 2$, and the observed resonance is from the intradoublet transition within the $|\pm2\rangle$ doublet (alternatively: the $|2^{sym}/2^{anti}\rangle$ doublet) (Hagen 1982b). The simulation of the experimental

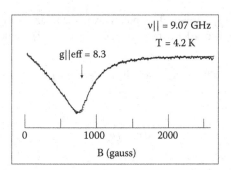

FIGURE 12.8 Simulation, based on an analytical expression, of parallel-mode EPR of an $S = 2$ system. The spectrum is from the hydrated ferrous ion $Fe^{II}(H_2O)_6$.

spectrum "goes through the dots," which is spectroscopy lingo for an excellent fit. The simulation is obtained in a split second because the analytical expressions in Equations 12.14 and 12.15 are extremely economic in terms of CPU time. Three fitting parameters are required: $g_\parallel^{eff} = 8.3$; $b_{0\parallel} = 400$ gauss; $b_{r_\parallel} = 450$ gauss. The slightly broader and high-field shifted normal-mode spectrum can be obtained in another split second with Equation 12.16 in the simulator.

13 Black Box Experiments

There is spectroscopy and there is spectrometry, and the difference between the two can range from imperceptibly small to unbridgeably large. Up to this point we have been dealing with the art of bioEPR spectroscopy; let us now consider the craftsmanship of bioEPR spectrometry. As a first attempt to define their difference, we turn to the analogy of the Uv-Vis spectrometer: optical spectra may be scrutinized to learn about the electronic structure of proteins, but it is also possible to fix the monochromator at 280 nm and to fix the experimenter's mind at determining the number of milligram per milliliter protein while waving away any thought on low-energy transitions in conjugated systems of aromatic amino-acid side groups. The latter spectrometric experiment is very much more common than the former spectroscopic one and, although it may be intellectually very much less challenging, its practical impact (protein purification, food analysis, clinical assays) is competitive, to say the least. The milligram per milliliter determination does not require any background in quantum mechanics, but for its proper execution it may be helpful to know that nonproteinaceous compounds (e.g., RNA) may interfere, and that optical photomultipliers saturate at absorptions somewhere above unity.

The EPR spectrometric equivalent of the 280-nm measurement would be to "blindly" monitor the amplitude of any feature in an EPR spectrum as the flag of some state of matter, using a fixed set of spectrometer settings. This black box type of relative concentration monitoring with EPR, in which we do not care about detailed interpretation of spectra as long as they can be assigned to a (bio)chemical species, allows in principle for considerable sloppiness on the part of the spectrometer operator: a signal can be overmodulated, partially saturated, or deformed by filtering as long as the spectrometer settings are kept constant such that the suboptimal measuring conditions are reproducible. This implies that we can go for maximal signal-to-noise (of a deformed signal), that is, for an improved detection limit in EPR spectrometry compared to optimized conditions for EPR spectroscopy. Thus, monitoring an EPR amplitude from a dilute sample as a function of some external thermodynamic or kinetic parameter can provide quite valuable biochemical information even when the spectra are not particularly "publication quality." As always, however, there is a catch: in multicenter molecules subject to changing external conditions, the intrinsic properties (e.g., the spin-lattice relaxation rate) of the paramagnet under monitoring may be influenced by another paramagnet nearby, and so the apparent amplitude of a partially saturated signal may change even if the spin concentration does not. For these situations craftsmanship of a well-informed operator is obviously at least as important as in regular detection of high-quality spectra.

13.1 EPR-MONITORED BINDING EXPERIMENTS

The stoichiometric binding of compound A to compound B (which is identical to the binding of compound B to compound A)

$$A + B \underset{k_{-1}}{\overset{k_1}{\rightleftharpoons}} AB \qquad (13.1)$$

is characterized by the equilibrium constant (or association constant) K_a of their reaction:

$$K_a = \frac{[AB]}{[A][B]} = \frac{k_1}{k_{-1}} = \frac{1}{K_d} \qquad (13.2)$$

in which K_d is the dissociation constant. In introductory textbooks on biochemistry, this subject usually appears in the chapter on signal transduction, and the chemical equilibrium is written in terms of the binding of a ligand L to a receptor protein R

$$R + L \underset{k_{-1}}{\overset{k_1}{\rightleftharpoons}} RL \qquad (13.3)$$

that is, in terms of a saturable binding site on a big biomolecule in the presence of excess of a much smaller organic (a hormone). The binding is then characterized by means of a straight-line plot of [RL]/[L] versus [RL] known as the *Scatchard analysis*:

$$\frac{[RL]}{[L]} = \frac{1}{K_d} \left(B_{max} - [RL] \right) \qquad (13.4)$$

in which $B_{max} = [R] + [RL]$ is the total number of binding sites (usually equal to unity) and $-1/K_d$ is the slope of the line.

The Scatchard formalism can of course be applied to the binding of any small molecule to any biomacromolecule, such as the binding of a substrate or inhibitor to an enzyme, or the binding of a metal ion to an apoprotein. In receptor research, the determination of K_d typically requires labeling of the substrate by radioactivity or by fluorescence. However, we might just as well choose paramagnetism as the label, and this then makes the EPR spectrometer the detector for the determination of binding equilibria. The Scatchard plot in Equation 13.4 has two experimental observables: [L] and [RL], and so we must find ways to determine these quantities from EPR spectra.

An obvious and straightforward experiment would be to measure the binding of a metal ion to a protein by monitoring the EPR signal of the free ion (L) and that of the bound ion (RL). In practice, things are usually a bit less straightforward. When a Cu^{II} ion is added to a protein, the Cu–protein complex (RL) usually has a characteristic and well-defined EPR spectrum, but the aqueous copper ion tends to aggregate affording a weak and broad signal that may easily go undetected, which means that we have no direct measure of [L]. If the used buffer has coordinative capacity, then a clear Cu-buffer spectrum may be found (cf. the Cu-Tris buffer signal in Chapter 5, Figure 5.7), but now spectral overlap with the Cu–protein signal may complicate the analysis. If [L] (i.e., "free" copper) cannot be directly measured, its value can be deduced from the known total amount L_{total} of metal ion added to the solution:

$$[L_{free}] = [L_{total}] - [RL] \qquad (13.5)$$

The Mn^{II} ion in solution gives a simple and sharp isotropic sextet spectrum ($I = 5/2$), but binding to a protein produces a broad, anisotropic spectrum (cf. Chapter 5, Figure 5.12) that may be difficult to detect at room temperature. So, now we do see L but we do not see RL. In this case, we can deduce [RL] from the decrease of the EPR signal from free Mn^{II} upon addition of protein.

The literature contains a rich variety of examples of EPR-monitored binding studies, from which we randomly cite only a few here: binding of copper to a prion protein monitored on the low-temperature Cu–protein EPR signal (Aronoff–Spencer et al. 2000); binding of Mn^{II} to a 15-mer DNA quadrupex (Marathias et al. 1996) or to the enzyme 3-OH-3-methylglutaryl-CoA lyase (Roberts and Miziorko 1997) monitored on the free Mn EPR signal; displacement of vanadyl by ferrous ion binding to ferritin monitored on the VO^{2+}-protein EPR signal (Chasteen and Theil 1982); binding of vanadate to ATP synthase by monitoring a change in the EPR spectrum of a spin label bound to the protein (Coan et al. 1996); binding of ligand to the acyl-CoA-binding protein monitored on the EPR spectrum of the ligand, that is, an esterified spin-labeled fatty acid (Rosendal et al. 1993).

The general message here is that EPR-monitored binding studies frequently require a bit of wet biochemical ingenuity in addition to EPR spectroscopic skills. Note in particular that binding may be more complex than suggested by the simplicity of Equation 13.1, for example, involving nonspecific binding and/or binding to multiple sites with different affinities.

13.2 EPR MONITORING OF REDOX STATES

A redox reaction is a special case of the equilibrium reaction of $A + B$ in Equation 13.1: B is now a reducible group in a biomolecule with an EPR spectrum either in its oxidized or in its reduced state (or both), and A is now an "electron" or a "pair of electrons," that is, reducing equivalents provided by a natural redox partner (a reductive substrate, a coenzyme such as NADH, a protein partner such as cytochrome c), or by a chemical reductant (dithionite), or even by a solid electrode:

$$B^{x+} + ne^- \rightleftharpoons B^{(x-n)+} \qquad (13.6)$$

Of course free electrons do not exist in a regular laboratory, so Equation 13.6 is a theoretical half reaction of the full equilibrium:

$$A_{red} + B_{ox} \rightleftharpoons A_{ox} + B_{red} \qquad (13.7)$$

Usually we know the properties of the electron donor, and we want to use EPR spectroscopy to determine those of the acceptor only. So we write down the Nernst equation for a single redox pair (replacing B with X):

$$E = E^0 + Q \ln\left([X_{ox}]/[X_{red}]\right) \tag{13.8}$$

in which, at a temperature of 25°C ($T = 298.15$ K), and with the gas constant $R = 8.31441$ J$\times$mol$^{-1}\times$K^{-1} and the Faraday constant $F = 96484.6$ C$\times$mol^{-1}:

$$Q = (RT)/(nF) = (38.92n)^{-1} \tag{13.9}$$

with n being the number of electrons transferred. Equation 13.8 is also written at 25°C as

$$E = E^0 + \frac{0.059}{n} \log\left(\frac{[X_{ox}]}{[X_{red}]}\right) \tag{13.10}$$

Equation 13.8 can be rearranged to

$$[X_{red}] = \frac{[X_{ox}] + [X_{red}]}{1 + \exp((E - E^0)/Q)} \tag{13.11}$$

which is a useful form for EPR analysis because it gives the concentration of reduced molecules in terms of the total concentration of X. Now suppose that X_{ox} is a closed-shell system ($S = 0$) and is a one-electron acceptor so that X_{red} is a half-integer paramagnet; then Equation 13.11 can be rewritten in terms of an EPR amplitude I as a function of the maximal amplitude I_{max} for a fully reduced preparation as

$$I_{red} = I_{max}/(1 + \exp((E - E^0)/Q)) \tag{13.12}$$

In other words, if we subject a homogeneous solution of X to an electrochemical potential E, then the amplitude of the EPR spectrum from this (possibly frozen) solution will be given by Equation 13.12. If we make samples for several different values of E, then their collective EPR amplitudes make a graph of I_{red} versus E that will define the value of the unknown E^0, the standard reduction potential (biochemists call this the midpoint potential) of the X_{red}/X_{ox} couple.

An example is given in Figure 13.1 for the reduction of the [2Fe–2S](Cys)$_4$ cluster in a ferredoxin, and the [2Fe–2S](Cys)$_2$(His)$_2$ cluster in an oxygenase enzyme, two proteins that are part of a three-protein chain for the oxidation of the herbicide "dicamba" (Chakraborty et al. 2005). The E^0's are read out as E-values corresponding to half-maximal EPR amplitude. The E^0 of the [2Fe–2S] cluster with two histidine ligands (a so-called *Rieske cluster*) in the enzyme is seen to have a less negative value of -0.02 V than the $E^0 = -0.17$ V of the all-cysteine cluster in the ferredoxin, which is consistent with the ferredoxin being the physiological reductant of the enzyme.

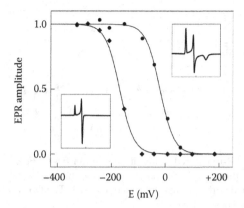

FIGURE 13.1 An EPR-monitored redox titration of two [2Fe–2S] clusters. A 2Fe cluster in a ferredoxin ($E^0 = -170$ mV) and one in an oxygenase enzyme ($E^0 = -20$ mV) from *Pseudomonas maltophilia* were each titrated with dithionite in the presence of a mediator mix. Each point is the EPR amplitude from an individual sample drawn at the indicated solution E-value. The fit is based on Equation 13.12. (Data from Chakraborty et al. 2005.)

For a redox center with an EPR signal in its oxidized state (e.g., a Cu^{II}/Cu^{I} couple or an Fe^{III}/Fe^{II} couple), the maximal intensity would be on the right side of the graph and would fit the equation

$$I_{ox} = I_{max} / (1 + \exp((E^0 - E) / Q)) \tag{13.13}$$

(Note the exchange of E and E^0 compared to Equation 13.12.) There are also quite a number of systems that can occur in three subsequent oxidation states, usually with the intermediate state being half-integer spin, such as $Co^{III}/Co^{II}/Co^{I}$, or $Mo^{VI}/Mo^{V}/Mo^{IV}$. In a reductive titration, the EPR intensity initially increases, and subsequently decreases again according to

$$I_{intermediate} = I_{max} / (1 + \exp((E - E_1^0) / Q) + \exp((E_2^0 - E) / Q)) \tag{13.14}$$

An example is given in Figure 13.2 for the dinuclear iron center in ferritin, which is $S = 1/2$ in the mixed-valence [Fe^{III}-Fe^{II}] state (Tatur and Hagen 2005). The two subsequent reduction steps [Fe^{III}-Fe^{III}]/[Fe^{III}-Fe^{II}] and [Fe^{III}-Fe^{II}]/[Fe^{II}-Fe^{II}] have $E_1^0 = +0.21$ V and $E_2^0 = +0.05$ V, which implies that the maximal EPR intensity never reaches the 100% level. In other words, it is impossible in an equilibrium redox titration to prepare the protein in a state in which all the molecules have their dinuclear cluster in the [Fe^{III}/Fe^{II}] mixed-valence state. This maximum further reduces for systems in which the two E^0's are closer in value: for $E_1^0 = E_2^0$, the maximum is only 33%. Many organic compounds are electron-*pair* donors/acceptors, which means that they can undergo two subsequent one-electron redox reactions with $E_1^0 \ll E_2^0$, that is, the potentials have "crossed over" and the intermediate radical state has very low stability and, therefore, very low EPR signal intensity.

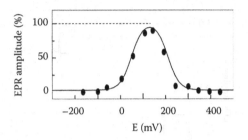

FIGURE 13.2 An EPR-monitored redox titration of an Fe–O–Fe cluster with three stable oxidation states. The dinuclear iron center (= +210 mV and = +50 mV) in *Pyrococcus furiosus* ferritin was titrated in the presence of a mediator mix. The fit is based on Equation 13.14. (Data from Tatur and Hagen 2005.)

To prepare EPR samples of proteins at a fixed redox potential requires a relatively simple setup schematically outlined in Figure 13.3. Because an EPR sample has a volume of 100–200 μL, we need circa 1.5 mL of anaerobic (cf. Chapter 3, Section 3.5) protein solution to collect data for a ten-point amplitude versus potential graph. The protein concentration may be significantly lower than that of a sample for spectroscopic analysis because we are interested only in the relative EPR amplitude for each sample recorded under conditions that maximize signal-to-noise ratio: a single 200 μL EPR

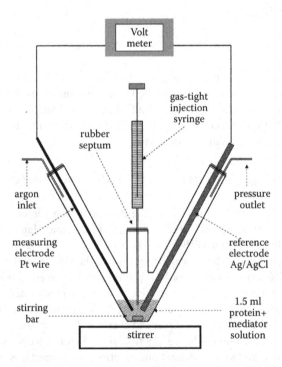

FIGURE 13.3 Schematic drawing of an electrochemical cell for mediated redox titrations of metalloproteins.

spectroscopy sample may thus be diluted tenfold to produce ten spectrometry samples. The solution potential is set by repeated addition of *substoichiometric* amounts of anaerobic reductant (e.g., dithionite) by means of a gas-tight injection syringe under continuous stirring of the protein solution. The solution potential is measured by means of a platinum electrode with respect to a reference electrode, for example, Ag|AgCl (i.e., a silver wire in a saturated KCl solution with E^0 = +197 mV at 25°C versus the normal hydrogen electrode) connected to a simple volt meter (a 5€ multimeter from the local do-it-yourself shop will do). After each addition of reductant, we wait circa 5–10 min until equilibrium is reached as evidenced from a stable potential reading (<5 mV/minute drift), and then we take a sample with a second gas-tight injection syringe to be transferred to an anaerobic EPR tube waiting on the gas manifold (Chapter 3, Section 3.5), and to be frozen immediately.

To reach redox equilibrium between a protein and a reductant such as dithionite, and particularly between a protein and a solid platinum wire, it can take very much longer than the few minutes we are prepared to wait before drawing a sample. In order to be able to complete a titration in a few hours instead of many days, we must drastically increase the approach-to-equilibrium rate by the addition of redox mediators to the protein solution (Dutton 1978). Redox mediators are organic dyes that tend to react rapidly with the chemical reductant, with the protein, and with the platinum electrode. However, a mediator is a redox compound itself, so it has its own standard reduction potential E^0, and it is only capable of stimulating the approach to equilibrium in a limited potential range around its E^0 value. Most mediators are two-electron donors/acceptors ($n = 2$), which means that at 29.5 mV away from their E^0, the ratio the two forms (*ox* and *red*) is ten, and at 59 mV from E^0, it is 100 (Equation 13.10). At potentials further away from E^0, the concentration of one of the forms becomes so low that the mediating capacity becomes negligible. This means that, if we want to carry out a redox titration of a protein over a potential range of, say, +400 to −400 mV (cf. Figures 13.1 and 13.2; also recall the biological potential window in Chapter 3, Figure 3.7), we must add a whole collection of redox mediators each with a different E^0, preferably at equidistant steps over the −400 to +400 mV range. Here are a few practical notes:

i. An example of a "mediator mix" that is routinely used in my lab (Pierik et al. 1993), runs from methyl viologen (MV: $E^0 \approx -0.45$ V) to tetramethyl-*p*-phenylenediamine (TMPD: $E^0 \approx +0.26$ V), namely: methyl viologen, benzyl viologen (−0.36 V), neutral red (−0.32 V), safranin O (−0.28 V), phenosafranin (−0.25 V), anthraquinone-2-sulfonate (−0.22 V), 2-hydroxy-1,4-naphtoquinone (−0.15 V), indigo disulfonate (−0.12 V), resorufin (−0.05 V), methylene blue (+0.01 V), phenazine ethosulfate (+0.06 V), 2,6-dichloroindophenol (+0.22 V), and N,N,N′,N′-tetramethyl-*p*-phenylenediamine. See Clark (1960) for these and many more E^0 values.

ii. The action of a mediator is catalytic, so its concentration can be substoichiometric with respect to the protein; however, the difference should not be more than a factor of ten for practical equilibration times.

iii. The setup described in Figure 13.3 can be used for the determination of E^0 values down to approximately −400 mV. If a protein has a prosthetic group with $E^0 < -400$ mV, then it may be increasingly difficult to attain equilibrium due to leaking in of traces of air. The titration cell then has to be transferred in its entirety to an anaerobic glove box.

iv. Mediators are strongly colored and they have a tendency to "stick" to proteins. Once a protein has been used in a redox titration, it cannot usually be recovered for other purposes.

v. Most mediators are $n = 2$ compounds, and they have very low radical concentration at potentials around their E^0-value, but at the extremes of the mediator-mix collection we find bona fide $n = 1$ compounds (MV, TMPD) that produce stoichiometric amounts of radical upon one-electron reduction. The EPR signals of these radicals are typically much sharper than the signals from metal centers, so interference can be avoided by measuring the metal spectrum at a position outside the radical spectral range.

vi. A possible alternative to mediated redox titration is the titration of an oxidoreductase enzyme with its natural substrate. For an isolated enzyme, in the absence of its natural redox partner (e.g., a ferredoxin, a cytochrome), enzyme turnover cannot proceed. The potential is then set by addition of a specific ratio of substrate over product. For example, addition of equal amounts of substrate and product will poise the potential at a value equal to the E^0 of the substrate/product couple. No mediators are required because the substrate rapidly equilibrates with the enzyme. The potential cannot be read out but is deduced from the substrate–product concentrations and their known E^0-value. Of course, this approach only works when the substrate's E^0 is rather close in value to that of the prosthetic group in the enzyme.

We close this section with a note on the influence of pH on reduction potentials. Many redox reactions are pH-dependent, which can be understood with reference to the simple model in Figure 13.4, in which a redox compound in its oxidized state has a pK_a for proton dissociation that is different from (i.e., lower than) the corresponding value for its reduced state: the positive charge of X_{ox} is higher than that of X_{red}, so it is more difficult for X_{ox} to accept a proton (i.e., its pK_a is lower). The $E^0(pH)$ is now

$$E^0 = E^0_{low\ pH} + Q \ln\left(\frac{[H^+] + K_{red}}{[H^+] + K_{ox}}\right) \tag{13.15}$$

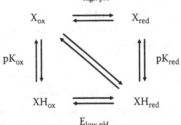

FIGURE 13.4 Four-state scheme for a redox couple with a single protonatable site.

that defines three ranges: below pH = pK_{ox} the E^0 is constant (maximum value); between pK_{ox} < pH < pK_{red} the E^0 decreases with 59/n mV per pH-unit increase; above pH = pK_{red} the E^0 is again constant (minimum value). Equation 13.15 becomes more complex when more than one protonatable groups are involved (Clark 1960).

Note, however, that the −59/n mV change per pH-unit is seldom found for prosthetic groups in proteins because association of protons is usually not directly on the coordination complex (which could result in loss of the metal) but rather on a nearby (or not-so-nearby) amino-acid side chain. So, the change can be anywhere between 0 and −59/n mV. This information can be quite valuable for an understanding of the mechanism of action of the metalloprotein, but it does mean that we have to carry out EPR-monitored redox titrations at several different pH-values.

The pH-dependence is of particular relevance for groups that occur in three subsequent oxidation states because the two reduction potentials E_1^0 and E_2^0 in Equation 13.14 in general have different pH dependence. For example, the paramagnetic W^V state of the tungsto-enzyme DMSO reductase affords an EPR signal with a maximal spin count of 40% of protein concentration at pH = 5 when $E_1^0 - E_2^0 \approx +10$ mV, whereas at pH = 8 no signal is detected at all because $E_1^0 \ll E_2^0$ (Hagedoorn et al. 2003).

13.3   EPR MONITORED KINETICS

In principle, EPR spectrometry is well suited as a method to monitor kinetic events; however, in practice, the time required to tune the spectrometer, and its intrinsically low sensitivity compared to fluorescence or light-absorption spectrometry, affect its competitiveness. Relatively slow reactions on the timescale of minutes, such as the decomposition of the DMPO-superoxide adduct and the subsequent formation of the hydroxyl radical adduct (cf. Pou et al. 1989) are readily followed, either as the first-order disappearance of the DMPO/•OOH signal

$$I_{decay} = I_{max} \exp(-kt) \tag{13.16}$$

or as the quasi first-order formation of the DMPO/•OH signal

$$I_{formation} = I_{max} (1 - \exp(-kt)) \tag{13.17}$$

but real-time monitoring of truly fast events is usually limited to reactions that can be triggered by a light pulse, such as photosynthesis-related events in the solid state.

On the contrary, bioEPR spectroscopy has gained a fine reputation in the off-line study of reaction intermediates produced by the method of rapid mixing plus rapid freezing originally developed by the late Bob Bray and his collaborators (Palmer et al. 1964). The EPR spectroscopic part of this approach is not different from that on regular samples, but the sample preparation part is much more demanding. The principle is quite simple: two reactants, for example, a metalloprotein and its substrate, are efficiently mixed and subsequently given a certain time to react in a tube of variable length (corresponding to variable reaction time) following the mixing chamber, until the mixture reaches the end of the tube from which it is injected in a cryogenic liquid to stop the reaction. Then the sample is collected in an EPR tube for study in the EPR spectrometer.

The hard part is in the "efficient mixing" and in the "sample collection." To mix within a time to resolve enzyme reactivity requires special mixing chambers connected to a rapid mixing machine, that is, a setup that can eject the content of two injection syringes in a fraction of a second into a mixing chamber by means of a ram mechanism. Commercially available equipment (order-of-magnitude price tag: 100 k€) can reach a "dead time" (i.e., the minimum time required to obtain efficient mixing) of 5–10 ms. Part of the requirements to reach this specification is the necessity to spray the reaction solution in a cold liquid (usually isopentane cooled by liquid nitrogen), which results in a fine "snow" of ice flakes that have the unfortunate tendency to charge up electrostatically and therefore to resist being pushed into an EPR tube (all handling in liquid isopentane). The degree of packing is now a new EPR sample parameter, and not always a very cooperative one.

In a recent development, Simon de Vries has pushed the dead time of the method well into the microsecond range using a new type of mixing chamber (called a tangential micro-mixer) and employing a liquid-nitrogen-cooled, rapidly revolving tungsten-coated drum as the reaction stopper (Cherepanov and De Vries 2004). There are plenty of enzymes out there that are actually so fast as to require such dead times for the resolution of their kinetics, and so, for go-getting characters, there is still much to be discovered in kinetic bioEPR spectroscopy.

In conclusion, rapid-mixing/rapid-freezing EPR is a wonderful technique to obtain unique molecular structural information on biochemical reaction intermediates with high time resolution, but it is also experimentally sufficiently involved that one should either build up a dedicated lab with dedicated operators or turn to one of the existing groups that have the equipment and, especially, the developed skills to do these experiments. Be prepared to provide at least an order of magnitude more sample than required for a static EPR experiment.

Finally, as a poor man's alternative, consider the possibility to slow down the reaction kinetics by running a reaction at $\leq 0°C$ temperatures (especially by employing enzymes from hyperthermophilic species) so that the mixing may be done in seconds (by hand!), and let us then hope that the kinetic mechanism under these "nonphysiological" conditions still bears relevance to the natural biology.

13.4 EPR OF WHOLE CELLS AND ORGANELLES

Thus far we tacitly assumed that our EPR tubes were filled with pure, or at least to some extent purified, biomolecular preparations. However, what will we find when we try to measure, for example, whole cells? In other words, how complex can a sample actually be without us losing track of all the overlaying signals? What would be the dynamic range of signal amplitudes that we can resolve from a single sample?

Measuring whole cells, or perhaps purified organelles from whole eukaryotic cells, for example, mitochondria, goes back to the very first days of bioEPR spectroscopy (Beinert and Lee 1961) and has since then over and over again proven to be useful for the particular purpose of studying respiratory chains, that is, the set of redox enzymes that form the heart of the bioenergetic machinery and that, for this reason, typically occur in high concentration in the cell. An example is the spectrum of whole heart tissue in Figure 13.5: the many overlapping spectral components are reasonably well resolved, and they have been assigned to a wide range of paramagnetic centers in the respiratory complexes of the rat (Van der Kraaij et al. 1989). The detailed assignment is of course not obvious from inspection of this single spectrum only but is the result of many years of study in multiple laboratories on whole cells, mitochondria, submitochondrial particles (i.e., mitochondrial membranes holding the respiratory complexes), and purified respiratory proteins under a variety of external conditions such as of redox potential, pH, incubation with substrates (NADH, succinate), etc., and a variety of spectroscopic conditions such as temperature, microwave power, and microwave frequency.

In the EPR of mammalian cells, we do not see much in addition to the signals from the respiratory complexes. The enzyme aconitase from the citric-acid cycle can be detected, and also the protein "cytoplasmic aconitase," later identified as the mRNA translation regulatory factor iron regulatory protein IRP-1, which actually started its career in biochemistry as an EPR signal that could not be assigned to the respiratory chain (Kennedy et al. 1992).

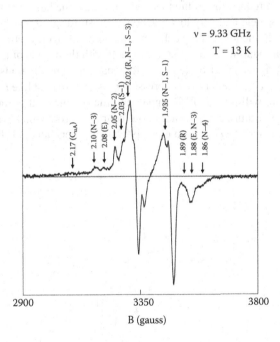

FIGURE 13.5 Whole eukaryotic-cell EPR. A rat heart was frozen in liquid nitrogen and ground to a fine powder. The EPR spectrum shows signals from prosthetic groups in respiratory chain complexes: N1–N4 (iron–sulfur clusters in NADH dehydrogenase); S1, S3 (iron–sulfur clusters in succinate dehydrogenase); R (the Rieske iron–sulfur cluster in the bc_1 complex); Cu_A (the mixed-valence copper dimer in cytochrome c oxidase); E (the iron–sulfur cluster in electron-transfer flavoprotein dehydrogenase). (Data from Van der Kraaij et al. 1989.)

This limitation is not always easily accepted by nonspectroscopists. On more than one occasion I have witnessed enthusiastic scientists from a variety of backgrounds bursting into the lab with the message that whole-cell EPR had just been proven in the literature to be capable of monitoring "dangerous" radicals related to items of added human interest such as ischemic heart reperfusion damage or aging of flowers of high commercial value. Let us refrain from citing the original literature on these "discoveries" and just note that the radical signals always turned out to be coming from "healthy" naturally occurring flavin and quinone components of the respiratory chain.

This earth is populated by many different types of cells, and their respiratory chains come in some variety. To illustrate this variation, Figure 13.6 gives the spectrum of cells from the sulfate-reducing bacterium *Desulfovibrio vulgaris*. The spectrum is not nearly as complex as that of the rat heart in Figure 13.5. Here, we only see a radical and a single anisotropic signal that is essentially identical to the spectrum (in Chapter 6, Figure 6.3) from a [2Fe–2S] cluster plus flavin radical in purified adenosine phosphosulfate (APS) reductase. This is understandable when one knows that this bacterium respires sulfate instead of oxygen, and to do this it requires three enzymes: one to activate the sulfate into APS (this enzyme has no paramagnetic groups), one to reduce the APS to sulfite with release of AMP (the enzyme that we see in the EPR), and then a third enzyme to reduce the sulfite to sulfide (this enzyme has prosthetic groups that have peaks at low-field and very-low-potential iron–sulfur groups, which are oxidized and therefore EPR silent in resting cells; cf. Marritt and Hagen 1996).

The situation found in Figure 13.6 (i.e., a cell with thousands of proteins but only one or two EPR signals) is also exemplary for what one typically finds when a recombinant paramagnetic protein is overexpressed in a standard host like *E. coli*. The overexpressed protein will give an EPR signal, and the background of the host is hardly detectable. The literature contains numerous examples from which we randomly cite a few (Uhlmann et al. 1997; Gaudu et al. 1997; Gao-Sheridan et al. 1998).

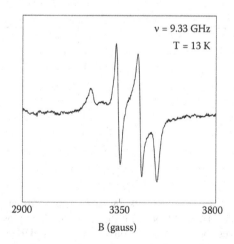

$\nu = 9.33$ GHz

$T = 13$ K

2900   3350   3800

B (gauss)

FIGURE 13.6 Whole bacterial-cell EPR. A frozen concentrated suspension of cells from the sulfate-reducing bacterium *Desulfovibrio vulgaris* gives an EPR spectrum with only a [2Fe–2S]$^{1+}$ signal and a flavin radical signal, both from adenosine phosphosulfate reductase.

14 Strategic Considerations

In this brief closing chapter we take a step backward from the knobs and the tubes and the equations to develop a few general thoughts on the nitty-gritty of how to approach a research problem in biomolecular EPR spectroscopy.

14.1 BIO-INTEGRATED BIOEPR

As a starting point let us now rephrase the position that was developed in the preface and the introductory Chapter 1 by means of a negation. "Building the ultimate beast" is a standing expression in the EPR community to describe microwave physics and engineering research activities to push the possibilities of EPR spectrometers beyond extant specifications. On the contrary, our goal is rather *not* to build the ultimate beast but to befriend the spectrometer as a thoroughly domesticated, useful animal.

I have assumed the reader to have a firm background in some part of the life sciences (including bioinspired synthetic chemistry and catalysis), from which a more than superficial interest in bioEPR has emerged. Our preferred position is one of having a problem (in the life sciences) in search of a method (EPR), rather than the other way around. In this scenario the spectrometer is second in rank to the sample. However, even with a cursory reading of this book it will be clear that, once the choice for EPR is made, some serious work has to be done. For example, whereas a meaningful biooptical measurement may not require any background beyond Beer's law, the EPR equivalent is not only much more complicated (cf. Chapter 6, Equation 6.4 and Chapter 8, Equation 8.27), but even mastering it does not provide sufficient knowledge to collect and interpret bioEPR data. We want to be specialists of sorts, meaning that we master the EPR field at a level that allows full harvesting of biorelevant results, but we should set out on our research in such a way that we do not get lost in search of the ultimate microwave beast and thus endanger our bioroots. We balance the two interests by their maximal mutual integration: in practice the EPR is, as it were, next to the pipette and the pcr block (you may replace the latter two by whatever your most common utensils happen to be). This position dictates specific requirements for, as well as elucidates the limitations of, spectroscopy.

The first and foremost conditio sine qua non for conceptual integration of the bio and the spectral is their actual physical integration. Ideally, the EPR lab should be as close as possible to the bio labs (maximally separated by a leisurely stroll) so that we can set up a bioEPR feedback loop. The outcome of putting a sample in an EPR spectrometer is never quite what one expects or would have wished for, and the reading of the spectra frequently directly translates into a desired action to modify the sample in a wet lab or to replace it altogether with a different sample. The sample can, for example, be insufficiently purified, too low (or occasionally too high) in concentration, in the wrong oxidation state, in a mixture of oxidation states, have improper viscosity, be denatured, contain too much dissolved oxygen, be incubated with improper reactant concentration and/or for improper time, not form a glassy state, be insufficiently anaerobic, have improper ionic strength, and so on. Sample adjustment (or replacement) may take several rounds and several types of wet chemistry, and the complexity of the logistics of the operation increases rapidly with the distance between the bio-lab and the spectro-lab.

14.2 TO BE ADVANCED OR NOT TO BE ADVANCED

In Europe we sometimes use the term "the exact sciences" when we mean "the natural sciences," presumably to express our hope and desire that our theories and results will turn out to be more accurate and to better stand the test of time than those from other fields of research. This does not necessarily keep us from regularly using very poorly defined notions such as "advanced EPR." In the EPR community, an experiment ran at 500 GHz is generally considered to be more advanced than the same experiment done at 10 GHz; however, this tacitly agreed-upon grading of "advancementness" is not necessarily obvious or acceptable from a life-science perspective. Even in the not-so-common case that a 500-GHz experiment on a biological sample does produce more than just noise, the resulting spectra are in fact usually easier to interpret than their 10-GHz counterparts. So, what notion do we want to get across when using the adjective "advanced" in relation to EPR spectroscopy? Its usage appears to have reached full swing with the 1989 appearance of the book edited by Arnold Hoff under the title *Advanced EPR: Applications in Biology and Biochemistry* (cf. Hagen 1989). This collection of chapters from different authors is rather dominated by contributions on double-resonance techniques and/or high-frequency experiments, which suggests that "advanced" is perhaps equivalent to instrumentally complex and expensive. Some scientific journals, such as *The Journal of the American Chemical Society,* have in their instructions to authors the directive not to use the word *novel* in the title of a paper because anything worth publishing has to be novel by definition. Should we not be equally picky about our use of the word *advanced*, realizing that any EPR experiment that does not in any way advance our knowledge is probably not worth mentioning? So, let us mentally replace "advanced" by "relevant," and let us set out to plan our experiments on the basis of their putative relevance to the scientific question that we want to address.

Double-resonance spectroscopy involves the use of two different sources of radiation. In the context of EPR, these usually are a microwave and a radiowave or (less common) a microwave and another microwave. The two combinations were originally called ENDOR (electron nuclear double resonance) and ELDOR (electron electron double resonance), but the development of many variations on this theme has led to a wide spectrum of derived techniques and associated acronyms, such as ESEEM (electron spin echo envelope modulation), which is a pulsed variant of ENDOR, or DEER (double electron electron spin resonance), which is a pulsed variant of ELDOR. The basic principle involves the saturation (partially or wholly) of an EPR absorption and the subsequent transfer of spin energy to a different absorption by means of the second radiation, leading to the detection of the difference signal. The requirement of saturability implies operation at close to liquid helium, or even lower, temperatures, which, combined with long experimentation times, produces a significant extra running-cost factor compared to regular EPR. Perhaps more importantly, the increased complexity of the instrumentation requires additional engineering, operation, and maintenance skills. Double-resonance equipment is usually not part of a life-science research operation but is typically found in specialized laboratories with dedicated staff. Theory and practical analysis of double-resonance spectra is not part of this book because the intended readership is an unfit audience. When a perceived need for double-resonance experiments arises in our research, we carefully do our checks and balances.

The main application of double resonance is the resolution of hyperfine and particularly superhyperfine interactions that are not extractable from regular EPR spectra because they are lost in the inhomogeneous line. The biological relevance is in otherwise unavailable detailed information on the electronic structure and the coordination of active sites and their interaction with reactants, such as enzyme substrates. To be well prepared, check off the items in the following list.

1. Try to imagine what you can expect to get out of a double-resonance experiment (look for atoms with nuclear spins in relevant structures) and preevaluate its biological relevance.
2. Consult a specialist on (i) the possibilities for the combination of your sample and double resonance, (ii) sample requirements, (iii) what instrument to use (specifically, what microwave frequency), (iv) time requirements, and (v) cost requirements.
3. Realize that you (or your sample) will be physically away from your lab and that there will probably be only limited facilities to "adjust" your sample in the ways we discussed in the frame of the bioEPR feedback loop earlier.
4. Never take sensitivity claims for granted; do not try sample concentrations below circa 1 mM.
5. Thoroughly check and characterize your sample in regular EPR.
6. Realize that $S = 1/2$ is OK, $S = n/2$ is usually very difficult, and $S = n$ is a no go.
7. Have a clear idea of what you want to "discover"; realize that your collaborating specialist is also in the publish-or-perish business.

Quite similar checklists can be set up for other forms of nominally considered "advanced" EPR, in particular for high-frequency/high-field EPR. Occasionally, not so very high frequencies may combine well with double resonance, but more typically, high-frequency EPR is either to resolve small g-anisotropy in $S = 1/2$ radicals or to simplify (or make possible at all) high-spin (including integer spin) EPR by moving into the strong-field regime.

Here is an additional checklist to prepare for high-frequency EPR experiments:

1. Try to imagine (e.g., by extrapolation of X-band simulations to high frequency) what you can expect to get out of a high-frequency experiment, and preevaluate its biological relevance.
2. Realize that chances to get a signal at all typically decrease with increasing frequency.

3. Check whether room-temperature sample loading is required (this can destroy your anaerobic sample).
4. Check how the high field is calibrated (this determines the accuracy of g-values).

A rather different type of "advanced" EPR concerns the detection of transient intermediates, which was briefly discussed in Chapter 13, Section 13.4. Here, the "advanced" (i.e., instrumentally complex and with its own price tag) refers to the sample preparation, not to the spectroscopy. The common denominator of the rapid-freezing ($\geq$ 5 ms) or even hyperquenching ($\ll$ 1 ms) kinetic equipment with double resonance and high frequency EPR is the requirement for a specialized laboratory and knowledge infrastructure. This makes detailed planning of experiments, including expert consultancy, equally important.

Another additional checklist to prepare for rapid freeze EPR experiments is as follows:

1. Try to imagine (e.g., from optically monitored stopped-flow experiments) what you can expect to get out of a rapid-freeze experiment, and preevaluate its biological relevance.
2. Characterize EPR-monitored kinetics down to the 0.5 s dead time in your own lab.
3. Decide that slowing down the kinetics by temperature lowering is not a reasonable alternative.
4. Assure ample (e.g., the equivalent of 20 regular samples) availability of concentrated sample to allow for 50% loss in mixing systems and for significantly more than one data point in a time plot.

14.3 FRIDAY AFTERNOON EXPERIMENT

This closing section is slightly more serious than its title suggests. EPR spectroscopy happens to be quite suitable for doing quick and dirty trial experiments, and I propose to explicitly add this type of activity to our strategic planning board. The Friday afternoon experiment is a standing expression in science for using a few spare hours to try something out of the ordinary, based on a wild, poorly-based hypothesis, and thus to probe a slim chance to open up a novel field, while the rest of the faculty has started socializing in preparation for the weekend. Rumor has it that many great discoveries originate in Friday afternoon experiments, but the reverse is not generally true, and our aims are accordingly modest.

Operation skills are a valuable asset in EPR spectroscopy, in particular where many bioEPR experiments require He-flow cooling, and costly helium does not stop evaporating while slow and clumsy operators are biding their time. The adrenaline-raising of a real experiment is less compatible with the leisurely pace required for self-practicing or teaching EPR to the novice; however, the learning experience can be equally effective when running experiments at room temperature or at 77 K in a small finger dewar filled with 50 mL of cheap liquid nitrogen. The strategic proposal then is to use part of your spectrometer time for cheap, playful, silly experiments on whatever compound you come across in your fridge, chemicals cabinet, local grocery shop, or what have you. This is a good way of acquiring operational skills, for building a mental library of spectroscopic model systems, building a butterfly collection of compounds suitable for teaching, for having fun, and, who knows, perhaps even for stumbling across a little gem of a discovery.

For example, solid copper acetate is not only a classic model for exchange interaction, it is also a cheap, stable, relatively innocent sample for use at room temperature and 77 K to try out or to teach triplet EPR, parallel-mode EPR, depopulation, line sharpening with decreasing temperature, forbidden transitions, hyperfine structure (in the forbidden transition), and contamination (with monomeric Cu^{II}). Another example would be to take just about any foodstuff (tea leaves are great) at room temperature and at 77 K to measure radicals, manganese, and iron in biological samples (and, in passing, to make the point that radicals are an intrinsic, healthy part of life). As yet another example, the cheap protein serum albumin (bovine or human) has a specific metal binding site at its N-terminal and can be used to practice or show EPR-monitored binding experiments by multiple additions of substoichiometric metal ion and subsequent measurement at 77K. The protein binds one Cu^{II} specifically and half a dozen Cu^{II} nonspecifically (i.e., two different Cu^{II} spectra). This list goes on forever, and for the truly inquisitive mind, the EPR experience offers infinite facets.

References

Abragam, A. and Pryce, M.H.L. 1951. Theory of the nuclear hyperfine structure of paramagnetic resonance spectra in crystals. *Proceedings of the Royal Society of London* A 205: 135–153.

Abragam, A. and Bleaney, B. 1970. *Electron Paramagnetic Resonance of Transition Ions.* Oxford: Clarendon Press.

Albracht, S.P.J. 1974. A low-cost cooling device for EPR measurements at 35 GHz down to 4.8°K. *Journal of Magnetic Resonance* 13: 299–303.

Albracht, S.P.J., Van Verseveld, H.W., Hagen, W.R., and Kalkman, M.L. 1980. A comparison of the respiratory chain in particles from *Paracoccus denitrificans* and bovine heart mitochondria by EPR spectroscopy. *Biochimica et Biophysica Acta* 593: 173–186.

Anderson, P.W. 1963. Theory of magnetic exchange interaction: exchange in insulators and semiconductors. In *Solid State Physics Volume 14*, eds. F. Seitz and D. Turnbull. New York: Academic Press.

Antholine, W.E., Hanna, P.M., and McMillan, D.R. 1993. Low frequency EPR of *Pseudomonas aeruginosa* azurin; analysis of ligand superhyperfine structure from a type 1 copper site. *Biophysical Journal* 64: 267–272.

Arendsen, A.F., Verhagen, M.F.J.M., Wolbert, R.B.G., Pierik, A.J., Stams, A.J.M., Jetten, M.S.M., and Hagen, W.R. 1993. The dissimilatory sulfite reductase from *Desulfosarcina variabilis* is a desulforubidin containing uncoupled metalated siroheme and S = 9/2 iron-sulfur clusters. *Biochemistry* 32: 10323–10330.

Arendsen, A.F., Hadden, J., Card, G., McAlpine, A.S., Bailey, S., Zaitsev, V., Duke, E.H.M., Lindley, P.F., Kröckel, M., Traitwein, A.X., Feiters, M.C., Charnock, J.M., Garner, C.D., Marritt, S.J., Thomson, A.J., Kooter, I.M., Johnson, M.K., Van den Berg, W.A.M., Van Dongen, W.M.A.M., and Hagen, W.R. 1998. The "prismane" protein resolved: X-ray structure at 1.7 Å and multiple spectroscopy of two novel 4Fe clusters. *Journal of Biological Inorganic Chemistry* 3: 81–95.

Aronoff-Spencer, E., Burns, C.S., Avdievich, N.I., Gerfen, G.J., Peisach, J., Antholine, W.E., Ball, H.L., Cohen, F.E., Prusiner, S.B., and Millhauser, G.L. 2000. Identification of the Cu^{2+} binding sites in the N-terminal domain of the prion protein by EPR and CD spectroscopy. *Biochemistry* 39: 13760–13771.

Bagguley, D.M.S. and Griffith, J.H.E. 1947. Paramagnetic resonance and magnetic energy levels in chrome alum. *Nature* 160: 532–533.

Baker, J.M. and Bleaney, B. 1958. Paramagnetic resonance in some lanthanon ethyl sulphates. *Proceedings of the Royal Society London* A 245: 156–174.

Bastardis, R., Guihéry, N., and Suaud, N. 2007. Relation between double exchange and Heisenberg model spectra: application to the half-doped manganites. *Physical Review B* 75: 132403 (1–4).

Beinert, H, and Sands, R.H. 1960. Studies on succinic and DPNH dehydrogenase preparations by paramagnetic resonance (EPR) spectroscopy. *Biochemical and Biophysical Research Communications* 3: 41–46.

Beinert, H. and Lee, W. 1961. Evidence for a new type of iron containing electron carrier in mitochondria. *Biochemical and Biophysical Research Communications* 5: 40–45.

Bennett, J.E., Ingram, D.J.E., George, P., and Griffith, J.S. 1955. Paramagnetic resonance absorption of ferrihaemoglobin and ferrimyoglobin derivatives. *Nature* 176: 394–394.

Bennett, J.E. and Ingram, D.J.E. 1956. Analysis of crystalline haemoglobin derivatives by paramagnetic resonance. *Nature* 177: 275–276.

Beringer, R. and Castle, J.G. Jr. 1951. Microwave magnetic resonance spectrum of oxygen. *Physical Review* 81: 82–88.

Berrisford, J.M., Hounslow, A.M., Akerboom, J., Hagen, W.R., Brouns, S.J.J., Van der Oost, J., Murray, I.A., Blackburn, G.M., Waltho, J.P., Rice, D.W., and Baker, P.J. 2006. Evidence supporting a *cis*-enediol-based mechanism for *Pyrococcus furiosus* phosphoglucose isomerase. *Journal of Molecular Biology* 358: 1353–1366.

Bleaney, B. and Bowers, K.D. 1952. Anomalous paramagnetism of copper acetate. *Proceedings of the Royal Society of London* A 214: 451–465.

Bleaney, B., Bowers, K.D., and Pryce, M.H.L. 1955. Paramagnetic resonance in dilute copper salts III. Theory, and evaluation of the nuclear electric quadrupole moments of ^{63}Cu and ^{65}Cu. *Proceedings of the Royal Society of London* A 228: 166–174.

Bleaney, B. and Scovil, H.E.D. 1952. Paramagnetic resonance in praseodymium ethylsulphate. *Philosophical Magazine* 43: 999–1000.

Bleaney, B., Llewellyn, P.M., Pryce, M.H.L., and Hall. G.R. 1954. Nuclear spin of 241 Pu. *Philosophical Magazine* 45: 991–992.

Bogomolova, L.D., Jachkin, V.A., Lazukin, V.N., Pavlushkina, T.K., and Shmuckler, V.A. 1978. The electron paramagnetic resonance and optical spectra of copper and vanadium in phosphate glasses. *Journal of Non-Crystalline Solids* 28: 375–389.

Bramley, R. and Strach, S.J. 1983. Electron paramagnetic resonance spectroscopy at zero magnetic field. *Chemical Reviews* 83: 49–82.

Brautigam, D.L., Feinberg, B.A., Hoffman, B.M., Margoliash, E., Peisach, J., and Blumberg, W.E. 1977. Multiple low spin forms of the cytochrome c ferrihemochrome. EPR spectra of various eukaryotic and prokaryotic cytochromes c. *The Journal of Biological Chemistry* 252: 574–582.

Bray, R.C., Malmström, B.G., and Vänngård, T. 1959. The chemistry of xanthine oxidase. 5. Electron spin resonance of xanthine oxidase solutions. *The Biochemical Journal* 73: 193–197.

Brill, A.S. 1977. *Transition Metals in Biochemistry*. Berlin: Springer.

Carrington, A. and McLachlan, A.D. 1967. *Introduction to Magnetic Resonance; with Applications to Chemistry and Chemical Physics*. New York: Harper and Row. Reprinted as Harper International Edition, New York: Harper & Row, 1969.

Chakraborty, S., Behrens, M., Herman, P.L., Arendsen, A.F., Hagen, W.R., Carlson, D.L., Wang, X.-Z., and Weeks, D.P. 2005. A three-component dicamba *O*-demethylase from *Pseudomonas maltophilia*, strain DI-6: purification and characterization. *Archives of Biochemistry and Biophysics* 437: 20–28.

Chasteen, N.D. and Theil, E.C. 1982. Iron binding by horse spleen apoferritin; a vanadyl(IV) EPR spin probe study. *Journal of Biological Chemistry* 257: 7672–7677.

Cherepanov, A.V. and De Vries, S. 2004. Microsecond freeze-hyperquenching: development of a new ultrafast micro-mixing and sampling technology and application to enzyme catalysis. *Biochimica et Biophysica Acta* 1656: 1–31.

Clark, W.M. 1960. *Oxidation-Reduction Potentials of Organic Systems*. Baltimore: Waverly Press.

Coan, C., Scales, D.J., and Murphy, A.J. 1996. Oligovanadate binding to sarcoplasmatic reticulum ATPase; evidence for substrate analogue behaviour. *Journal of Biological Chemistry* 261: 10394–10403.

Czoch, R. and Francik, A. 1989. *Instrumental Effects in Homodyne Electron Paramagnetic Resonance Spectrometers*. Chichester: Ellis Horwood.

Davydov, R., Kuprin, S. Gräslund, A., and Ehrenberg, A. 1994. Electron paramagnetic resonance study of the mixed-valent diiron center in *Escherichia coli* ribonucleotide reductase produced by reduction of radical-free protein R2 at 77 K. *Journal of the American Chemical Society* 116: 11120–11128.

De Gennes, P.-G. 1960. Effects of double exchange in magnetic crystals. *Physical Review* 118: 141–154.

De Groot, M.S. and Van der Waals, J.H. 1960. Paramagnetic resonance in phosphorescent aromatic hydrocarbons. II: Determination of zero-field splitting from solution spectra. *Molecular Physics* 3: 190–200.

Ding, X.-Q., Bominaar, E.L., Bill, E., Winkler, H., Traitwein, A.X., Drüeke, S., Chaudhuri, P., and Wieghardt, K. 1990. Mössbauer and electron paramagnetic resonance study of the double exchange and Heisenberg-exchange interactions in a novel binuclear Fe(II/III) delocalized-valence compound. *Journal of Chemical Physics* 92: 178–186.

Dutton, P.L. 1978. Redox potentiometry: determination of midpoint potentials of oxidation-reduction components of biological electro-transfer systems. *Methods in Enzymology* 54: 411–435.

Ehrenberg, A., Eriksson, L.E.G., and Hyde, J.S. 1968. Electron-nuclear double resonance from flavin free radicals in NADPH dehydrogenase ("old yellow enzyme"). *Biochimica et Biophysica Acta* 167: 482–484.

Elmali, A. 2000. The magnetic super-exchange coupling in copper(II) acetate monohydrate and a redetermination of the crystal structure. *Turkish Journal of Physics* 24: 667–672.

Feher, G. 1956. Observation of nuclear magnetic resonances via the electron spin resonance line. *Physical Review* 103: 834–835.

Feher, E.R. 1964. Effect of uniaxial stress on the paramagnetic spectra of Mn^{3+} and Fe^{3+} in MgO. *Physical Review* 136: A145–A157.

Fittipaldi, M., Warmerdam, G.C.M., De Waal, E.C., Canters, G.W., Cavazzini, D., Rossi, G.L., Huber, M., and Groenen, E.J.J. 2006. Spin-density distribution in the copper site of azurin. *ChemPhysChem* 7: 1286–1293.

Fournel, A., Gambarelli, S., Guigliarelli, B., More, C., Asso, M., Chouteau, G., Hille, R., and Bertrand, P. 1998. Magnetic interactions between a $[4Fe-4S]^{1+}$ cluster and a flavin mononucleotide radical in the enzyme trimethylamine dehydrogenase: a high-field electron paramagnetic resonance study. *Journal of Chemical Physics* 109: 10905–10913.

Frankel, R.B., Papaefthymiou, G.C., and Watt, G.D. 1991. Variation of superparamagnetic properties with iron loading in mammalian feritin. *Hyperfine Interactions* 66: 71–82.

Freed, J.H. 1976. Theory of slow thumbling ESR spectra for nitroxides. In *Spin Labeling; Theory and Applications*, ed. L.J. Berliner. New York: Academic Press.

Fritz, J., Anderson, R., Fee, J., Palmer, G, Sands, R.H., Tsibris, J.C.M., Gunsalus, I.C., Orme-Johnson, W.H., and Beinert, H. 1971. The iron electron-nuclear double resonance (ENDOR) of two-iron ferredoxins from spinach, parsley, pig adrenal cortex and *Pseudomonas putida*. *Biochimica et Biophysica Acta* 253: 110–133.

Froncisz, W. and Hyde, J.S. 1980. Broadening by strains of lines in the g-parallel region of Cu^{2+} EPR spectra. *Journal of Chemical Physics* 73: 3123–3131.

Gaffney, B.J. and McConnell, H.M. 1974. The paramagnetic resonance spectra of spin labels in phospholipid membranes. *Journal of Magnetic Resonance* 16: 1–28.

Gao-Sheridan, H.S., Kempers, M.A., Khayat, R., Tilley, G.J., Armstrong, F.A., Sridhar, V., Prasad, G.S., Stout, C.D., and Burgess, B.K. 1998. A T14C variant of *Azotobacter vinelandii* ferredoxin I undergoes facile $[3Fe-4S]^0$ to $[4Fe-4S]^{2+}$ conversion *in vitro* but not *in vivo*. *The Journal of Biological Chemistry* 273: 33692–33701.

Gaudu, P., Moon, N., and Weiss, B. 1997. Regulation of the *soxRS* oxidative stress regulon. *The Journal of Biological Chemistry* 272: 5082–5086.

Gayda, J.-P., Bertrand, P., Deville, A., More, C., Roger, G., Gibson, J.F., and Cammack, R. 1979. Temperature dependence of the electronic spin-lattic relaxation time in a 2-iron-2-sulfur protein. *Biochimica et Biophysica Acta* 581: 15–26.

Gayda, J.-P., Bertrand, P., and Theodule, F.-X. 1982. Three-iron clusters in iron-sulfur proteins: An EPR study of the exchange interactions. *Journal of Chemical Physics* 77: 3387–3391.

Gloux, J., Gloux, P., Lamotte, B., Mouesca, J.-M., and Rius, G. 1994. The different [Fe₄S₄]³⁺ and [Fe₄S₄]⁺ species created by γ irradiation in single crystals of the (Et₄N)₂[Fe₄S₄(Sbenz)₄] model compounds: their EPR description and their biological significance. *Journal of the American Chemical Society* 116: 1953–1961.

Goldman, S.A., Bruno, G.V., and Freed, J.H. 1972. Estimating slow-motional rotational correlation times for nitroxides by electron spin resonance. *The Journal of Physical Chemistry* 76: 1858–1860.

Göppert-Mayer, M. 1931. Über Elementarakte mit zwei Quantensprüngen. *Annalen der Physik* 9: 273–294.

Griffith, J.S. 1971. Theory of E.P.R. in low-spin ferric haemoproteins. *Molecular Physics* 21: 135–139.

Griffith, O.H. and McConnell, H.M. 1966. A nitroxide–maleimide spin label. *Proceedings of the National Academy of Sciences of the USA* 55: 8–11.

Hagedoorn, P.L., Driessen, M.C.P.F., Van den Bosch, M., Landa, I., and Hagen, W.R. 1998. Hyperthermophilic redox chemistry: a re-evaluation. *FEBS Letters* 440: 311–314.

Hagedoorn, P.-L., Hagen, W.R., Stewart, L.J., Docrat, A., Bailey, S., and Garner, C.D. 2003. Redox characteristics of the tungsten DMSO reductase of *Rhodobacter capsulatus*. *FEBS Letters* 555: 606–610.

Hagen, W.R. 1981. Dislocation strain broadening as a source of anisotropic linewidth and asymmetrical lineshape in the electron paramagnetic resonance spectrum of metalloproteins and related systems. *Journal of Magnetic Resonance* 44: 447–469.

Hagen, W.R. 1982a. Electron Paramagnetic Resonance of Metalloproteins, Ph.D. thesis, The University of Amsterdam.

Hagen, W.R. 1982b. EPR of non-Kramers doublets in biological systems; characterization of an S = 2 system in oxidized cytochrome *c* oxidase. *Biochimica et Biophysica Acta* 707: 82–98.

Hagen, W.R. 1989. g-Strain: inhomogeneous broadening in metalloprotein EPR. In *Advanced EPR: Applications in Biology and Biochemistry*, ed. A.J. Hoff. Amsterdam: Elsevier.

Hagen, W.R. 1992. EPR spectroscopy of iron-sulfur proteins. In *Advances in Inorganic Chemistry, vol 38: Iron-Sulfur Proteins*, eds. R. Cammack and A.G. Sykes. San Diego, CA: Academic Press.

Hagen, W.R. 2006. EPR spectroscopy as a probe of metal centres in biological systems. *Dalton Transactions* 2006: 4415–4434.

Hagen, W.R. 2007. Wide zero field interaction distributions in the high-spin EPR of metalloproteins. *Molecular Physics* 105: 2031–2039.

Hagen, W.R. and Albracht, S.P.J. 1982. Analysis of strain-induced EPR-line shapes and anisotropic spin-lattice relaxation in a [2Fe-2S] ferredoxin. *Biochimica et Biophysica Acta* 702: 61–71.

Hagen, W.R., Dunham, W.R., Sands, R.H., Shaw, R.W., and Beinert, H. 1984. Dual-mode EPR spectrometry of O₂-pulsed cytochrome *c* oxidase. *Biochimica et Biophysica Acta* 765: 399–402.

Hagen, W.R., Dunham, W.R., Johnson, M.K., and Fee, J.A. 1985a. Quarter field resonance and integer-spin/half-spin interaction in the EPR of *Thermus thermophilus* ferredoxin. Possible new fingerprints for three iron clusters. *Biochimica et Biophysica Acta* 828: 369–374.

Hagen, W.R., Eady, R.R., Dunham, W.R., and Haaker, H. 1985b. A novel S = 3/2 EPR signal associated with native Fe-proteins of nitrogenase. *FEBS Letters* 189: 250–254.

Hagen, W.R., Hearsen, D.O., Sands, R.H., and Dunham, W.R. 1985c. A statistical theory for powder EPR in distributed systems. *Journal of Magnetic Resonance* 61: 220–232.

Hagen, W.R., Wassink, H., Eady, R.R., Smith, B.E., and Haaker, H. 1987. Quantitative EPR of an S = 7/2 system in thionine-oxidized MoFe proteins of nitrogenase. *European Journal of Biochemistry* 169: 457–465.

Hagen, W.R., Hearsen, D.O., Harding, L.J., and Dunham, W.R. 1985d. Quantitative numerical analysis of g strain in the EPR of distributed systems and its importance for multi-center metalloproteins. *Journal of Magnetic Resonance* 61: 233–244.

Hagen, W.R., Vanoni, M.A., Rosenbaum, K., and Schnackerz, K.D. 2000. On the iron-sulfur clusters in the complex redox enzyme dihydropyrimidine dehydrogenase. *European Journal of Biochemistry* 267: 3640–3646.

Hanania, G.I.H., Irvine, D.H., Eaton, W.A., and George, P. 1967. Thermodynamic aspects of the potassium hexacyanoferrate(III)-(II) system. II. Reduction potentials. *Journal of Physical Chemistry* 71: 2022–2030.

Harbour, J.R. and Bolton, J.R. 1975. Superoxide formation in spinach chloroplasts: electron spin resonance detection by spin trapping. *Biochemical and Biophysical Research Communications* 64: 803–807.

Harris, G. 1966. Low spin ferric hemoglobin complexes. *Theoretica Chimica Acta* 5: 379–397.

Hearshen, D.O., Hagen, W.R., Sands, R.H., Grande, H.J., Crespi, H.L., Gunsalus, I.C., and Dunham, W.R. 1986. An analysis of g strain in the EPR of two [2Fe-2S] ferredoxins. Evidence for a protein rigidity model. *Journal of Magnetic Resonance* 69: 440–459.

Heering, H.A., Bulsink, Y.B.M., Hagen, W.R., and Meyer, T.E. 1995. Reversible super-reduction of the cubane $[4Fe-4S]^{(3+;2+;1+)}$ in the high-potential iron-sulfur protein under non-denaturing conditions; EPR spectroscopic and electrochemical studies. *European Journal of Biochemistry* 232: 811–817.

Heller, C. and McConnell, H.M. 1960. Radiation damage in organic crystals. II. Electron spin resonance of $(CO_2H)CH_2CH(CO_2H)$ in β-succinic acid. *The Journal of Chemical Physics* 32: 1535–1539.

Holliger, C., Schraa, G, Stupperich, E., Stams, A.J.M., and Zehnder, A.J.B. 1992. Evidence for the involvement of corrinoids and Factor F_{430} in the reductive dechlorination of 1,2-dichloroethane by *Methanosarcina barkeri*. *Journal of Bacteriology* 174: 4427–4434.

Holuj, F. 1966. The spin Hamiltonian and intensities of the EPR spectra originating from large zero-field effects on ^6S states. *Canadian Journal of Physics* 44: 503–508.

Hubbell, W.L. and McConnell, H.M. 1971. Molecular motion in spin-labeled phospholipids and membranes. *Journal of the American Chemical Society* 93: 314–326.

Hubbell, W.L., Gross, A., Langen, R., and Lietzow, M.A. 1998. Recent advances in site-directed spin labeling of proteins. *Current Opinion in Structural Biology* 8: 649–656.

Hyde, J.S. and Dalton, L.R. 1979. Saturation-transfer spectroscopy. In *Spin Labeling II; Theory and Applications*, ed. L.J. Berliner. New York: Academic Press.

Israeli, A., Patt, M., Oron, M., Samuni, A., Kohen, R., and Goldstein, S. 2005. Kinetics and mechanism of the comproportionation reaction between oxoammonium cation and hydroxylamine derived from cyclic nitrones. *Free Radical Biology and Medicine* 38: 317–324.

Isomoto, A., Watari, H., and Kotani. M. 1970. Dependence of EPR transition probability on magnetic field. *Journal of the Physical Society of Japan* 29: 1571–1577.

Janzen, E.G. 1971. Spin trapping. *Accounts of Chemical Research* 4: 31–40.

Janzen, E.G. 1995. Spin trapping. In *Bioradicals Detected by ESR Spectroscopy*, eds. H. Ohya-Nishiguchi and L. Parker. Basel: Birkhauser Verlag.

Jeschke, G. 2005. EPR techniques for studying radical enzymes. *Biochimica et Biophysica Acta* 1707: 91–102.

Johnston, T.S. and Hecht, H.G. 1965. An automatic fitting procedure for the determination of anisotropic g-tensors from EPR studies of powder samples. *Journal of Molecular Spectroscopy* 17: 98–107.

Kennedy, M.C., Mende-Mueller, L., Blondin, G.A., and Beinert, H. 1992. Purification and characterization of cytosolic aconitase from beef liver and its relationship to the iron-responsive element binding protein. *Proceedings of the National Academy of Sciences of the USA* 89: 11730–11734.

Kent, T.A., Huynh, B.H., and Münck, E. 1980. Iron-sulfur proteins: Spin-coupling model for three-iron clusters. *Proceedings of the National Academy of Sciences of the USA* 77: 6574–6576.

Kim, J., Darley, D.J., Buckel, W., and Pierik, A.J. 2008. An Allylic ketyl radical intermediate in clostridial amino-acid fermentation. *Nature* 452: 239–242.

Kirkpatrick, E.S., Müller, K.A., and Rubins, R.S. 1964. Strong axial electron paramagnetic resonance spectrum of Fe^{3+} in $SrTiO_3$ due to nearest-neighbor charge compensation. *Physical Review* 135: A86–A90.

Kneubühl, F.K. and Natterer, B. 1961. Paramagnetic resonance intensity of anisotropic substances and its influence on line shapes. *Helvetica Physica Acta* 34: 710–717.

Kochelaev, B.I. and Yablokov, Y.V. 1995. *The Beginning of Paramagnetic Resonance.* Singapore: World Scientific.

Lagerstedt, J.O., Budamagunta, M.S., Oda, M.N., and Voss, J.C. 2007. Electron paramagnetic resonance spectroscopy of site-directed spin labels reveals the structural heterogeneity in the N-terminal domain of ApoA-I in solution. *The Journal of Biological Chemistry* 282: 9143–9149.

Le Pape, L., Lamotte, B., Mouesca, J.-M., and Rius, G. 1997. Paramagnetic states of four iron–four sulfur clusters. 1. EPR single-crystal study of 3+ and 1+ clusters of an asymmetric model compound and general model for the interpretation of the g-tensors of these two redox states. *Journal of the American Chemical Society* 119: 9757–9770.

Lim, L.W., Shamala, N., Mathews, F.S., Steenkamp, D.J., Hamlin, R., and Xuong, N.H. 1986. Three-dimensional structure of the iron-sulfur flavoprotein trimethylamine dehydrogenase at 2.4-A resolution. *The Journal of Biological Chemistry* 261: 15140–15146.

Luis, F., Del Barco, E., Hernández, J.M., Remiro, E., Bartolomé, J., and Tejada, J. 1999. Resonant spin tunneling in small antiferromagnetic particles. *Physical Review B* 59: 11837–11846.

Lundin, A. and Aasa, R. 1973. A simple device to maintain temperatures in the range 4.2-100 K for EPR measurements. *Journal of Magnetic Resonance* 8: 70–73.

Malmström, B.G., Mosbach, R., and Vänngård, T. 1959. An electron spin resonance study of the state of copper in fungal laccase. *Nature* 183: 321–322.

Malmström, B.G., and Vänngård, T. 1960. Electron spin resonance of copper proteins and some model complexes. *Journal of Molecular Biology* 2: 118–124.

Marathias, V.M., Wang, K.Y., Kumar, S., Pham, T.Q., Swaminathan, S., and Bolton, P.H. 1996. Determination of the number and location of the manganese binding sites of DNA quadruplexes in solution by EPR and NMR in the presence and absence of thrombin. *Journal of Molecular Biology* 260: 378–394.

Marritt, S. and Hagen, W.R. 1996. Dissimilatory sulfite reductase revisited. The desulfoviridin molecule does contain 20 iron ions, extensively demetallated sirohaem, and an S = 9/2 iron-sulfur cluster. *European Journal of Biochemistry* 238: 724–727.

Mayhew, S.G. 1978. The redox potential of dithionite and SO_2^- from equilibrium reactions with flavodoxins, methyl viologen and hydrogen plus hydrogenase. *European Journal of Biochemistry* 85: 535–547.

McConnell, H.M. and Chesnut, D.B. 1958. Theory of isotropic hyperfine interactions in π-electron radicals. *The Journal of Chemical Physics* 28: 107–117.

Michel, F.M., Ehm, L., Antao, S.M., Lee, P.L., Chupas, P.J., Liu, G., Strongin, D.R., Schoonen, M.A.A., Philips, B.L., and Parise, J.B. 2007. The structure of ferrihydrite, a nanocrystalline material. *Science* 316: 1726–1729.

Moan, J. and Wold, E. 1979. Detection of singlet oxygen production by ESR. *Nature* 279: 450–451.

Moriya, T. 1960. Anisotropic superexchange interaction and weak ferromagnetism. *Physical Review* 120: 91–98.

Neese, F. 2003. Quantum chemical calculations of spectroscopic properties of metalloproteins and model compounds: EPR and Mössbauer properties. *Current Opinion in Chemical Biology* 7: 125–135.

Neese, F. 2006. Importance of direct spin-spin coupling and spin-flip excitations to the zero-field splittings of transition metal complexes: a case study. *Journal of the American Chemical Society* 128: 10213–10222.

Neese, F. 2007. Calculation of the zero-field splitting tensor on the basis of hybrid density functional and Hartree-Fock theory. *The Journal of Chemical Physics* 127: 164112/1–9.

Newman, D.J. and Urban, W. 1975. Interpretation of S-state ion E.P.R. spectra. *Advances in Physics* 24: 793–844.

Nordio, P.L. 1976. General magnetic resonance theory. In *Spin Labeling: Theory and Applications*, ed. L.J. Berliner. New York: Academic Press.

Oda, M.N., Forte, T.M., Ryan, R.O., and Voss, J.C. 2003. The C-terminal domain of apolipoprotein A-I contains a lipid-sensitive conformational trigger. *Nature Structural Biology* 10: 455–460.

Pake, G.E. and Estle, T.L. 1973. *The Physical Principles of Electron Paramagnetic Resonance*. Reading, MA: W.A. Benjamin Inc.

Palmer, G., Bray, R.C., and Beinert, H. 1964. Direct studies on the electron transfer sequence in xanthine oxidase by electron paramagentic resonance spectroscopy. *The Journal of Biological Chemistry* 239: 2657–2666.

Papaefthymiou, V., Girerd, J.-J., Moura, I., Moura, J.J.G., and Münck, E. 1987. Mössbauer study of *D. gigas* ferredoxin II and spi-coupling model for the Fe_3S_4 cluster with valence delocalization. *Journal of the American Chemical Society* 109: 4703–4710.

Pardi, L.A., Krzystek, J., Telser, J., and Brunel, L.-C. 2000. Multifrequency EPR spectra of molecular oxygen in solid air. *Journal of Magnetic Resonance* 146: 375–378.

Pierik, A.J. and Hagen, W.R. 1991. S = 9/2 EPR signals are evidence against coupling between the siroheme and the Fe/S cluster prosthetic groups in *Desulfovibrion vulgaris* (Hildenborough) dissimilatory sulfite reductase. *European Journal of Biochemistry* 195: 505–516.

Pierik, A.J., Hagen, W.R., Dunham, W.R., and Sands, R.H. 1992a. Multi-frequency EPR and high-resolution Mössbauer spectroscopy of a putative [6Fe-6S] prismane-cluster-containing protein from *Desulfovibrio vulgaris* (Hildenborough); characterization of a supercluster and superspin model protein. *European Journal of Biochemistry* 206: 705–719.

Pierik, A.J., Hagen, W.R., Redeker, J.S., Wolbert, R.B.G., Boersma, M., Verhagen, M.F.J.M., Grande, H.J., Veeger, C., Mutsaers, P.H.A., Sands, R.H., and Dunham, W.R. 1992b. Redox properties of the iron-sulfur clusters in activated Fe-hydrogenase from *Desulfovibrio vulgaris* (Hildenborough). *European Journal of Biochemistry* 209: 63–72.

Pierik, A.J., Wassink, H., Haaker, H., and Hagen, W.R. 1993. Redox properties and EPR spectroscopy of the P clusters of *Azotobacter vinelandii* MoFe protein. *European Journal of Biochemistry* 212: 51–61.

Pilbrow, J.R. 1969. Anisotropic transition probability factor in E.S.R. *Molecular Physics* 16: 307–309.

Pilbrow, J.R. 1990. *Transition Ion Electron Paramagnetic Resonance*. Oxford: Clarendon Press.

Poole, C.P. Jr. 1983. *Electron Spin Resonance: A Complehensive Treatise on Experimental Techniques,* 2nd edition. New York: John Wiley & Sons. Reprinted Mineola, New York: Dover Publications, 1996.

Pou, S., Hassett, D.J., Britigan, B.E., Cohen, M.S., and Rosen, G.M. 1989. Problems associated with spin trapping oxygen-centered free radicals in biological systems. *Analytical Biochemistry* 177: 1–6.

Priem, A., van Bentum, P.J.M., Hagen, W.R., and Reijerse, E.J. 2001. Estimation of higher-order magnetic spin interactions of Fe(III) and Gd(III) ions doped in alpha-alumina powder with multifrequency EPR. *Applied Magnetic Resonance* 21: 535–548.

Priem, A.H., Klaassen, A.A.K., Reijerse, E.J., Meyer, T.E., Luchinat, C., Capozzi, F., Dunham, W.R., and Hagen, W.R. 2005. EPR analysis of multiple forms of [4Fe-4S]$^{3+}$ clusters in HiPIPs. *Journal of Biological Inorganic Chemistry* 10: 417–424.

Resnick, D., Gilmore, K., Idzerda, Y.U., Klem, M., Smith, E., and Douglas, T. 2004. Modeling of the magnetic behaviour of γ-Fe$_2$O$_3$ nanoparticles mineralized in ferritin. *Journal of Applied Physics* 95: 7127–7129.

Roberts, J.R. and Miziorko, H.M. 1997. Evidence supporting a role for histidine-235 in cation binding to human 3-hydroxy-3-methylglutaryl-CoA lyase. *Biochemistry* 36: 7594–7600.

Rosendal, J., Ertbjerg, P., and Knudsen, J. 1993. Characterization of ligand binding to acyl-CoA-binding protein. *Biochemical Journal* 290: 321–326.

Sands, R.H. and Beinert, H. 1959. On the function of copper in cytochrome oxidase. *Biochemical and Biophysical Research Communications* 1: 175–178.

Shaffer, J.S., Farach, H.A., and Poole, C.P. Jr 1976. Electron-spin-resonance of manganese-doped spinel. *Physical Review* B 13: 1869–1875.

Sharma, R.R., Das, T.P., and Orbach, R. 1966. Zero-field splitting of S-state ions. I. Point-multipole model. *Physical Review* 149: 257–269.

Sharma, R.R., Das, T.P., and Orbach, R. 1967. Zero-field splitting of S-state ions. II. Overlap and covalency model. *Physical Review* 155: 338–352.

Sharma, R.R., Das, T.P., and Orbach, R. 1968. Zero-field splitting of S-state ions. III. Corrections to part I and II and application to distorted cubic crystals. *Physical Review* 171: 378–388.

Sjöberg, B.-M., Reichard, P., Gräslund, A., and Ehrenberg, A. 1978. The tyrosine free radical in ribonucleotide reductase from *Escherichia coli*. *The Journal of Biological Chemistry* 253: 6863–6865.

Smith, C.A., Anderson, B.F., Baker, H.M., and Baker, E.N. 1992. Metal substitution in transferrins: the crystal structure of human copper-lactoferrin at 2.1-Å resolution. *Biochemistry* 31: 4527–4533.

Smoukov, S.K., Telser, J., Nernat, B.A., Rife, C.L., Armstrong, R.N., and Hoffmann, B.M. 2002. EPR study of substrate binding to the Mn(II) active site of the bacterial antibiotic resistance enzyme FosA: a better way to examine Mn(II). *Journal of the American Chemical Society* 124: 2318–2326.

Stevens, K.W.H. 1952. Matrix elements and operator equivalents connected with the magnetic properties of rare earth ions. *Proceedings of the Physical Society* 65: 209–215.

Stevens, K.W.H. 1997. *Magnetic Ions in Crystals*. Princeton, NJ: Princeton University Press.

Stevenson, R.C., Dunham, W.R., Sands, R.H., Singer, T.P., and Beinert, H. 1986. Studies on the spin-spin interaction between flavin and iron-sulfur cluster in an iron-sulfur flavoprotein. *Biochimica et Biophysica Acta* 869: 81–88.

Tatur, J. and Hagen, W.R. 2005. The dinuclear iron-oxo ferroxidase center of *Pyrococcus furiosus* ferritin is a stable prosthetic group with unexpectedly high reduction potentials. *FEBS Letters* 579: 4729–4732.

Taylor, C.P.S. 1977. The EPR of low spin heme complexes. Relation of the t_{2g} hole model to the directional properties of the g tensor, and a new method for calculating the ligand field parameters. *Biochimica et Biophysica Acta* 491: 137–149.

Teixeira, M., Campos, A.P., Aguilar, A.P., Costa, H.S., Santos, H., Turner, D.L., and Xavier, A.V. 1993. Pitfalls in assigning heme axial coordination by EPR. c-Type cytochromes with atypical Met-His ligation. *FEBS Letters* 317: 233–236.

Thomas, D.D., Dalton, L.R., and Hyde, J.S. 1976. Rotational diffusion studied by passage saturation transfer electron paramagnetic resonance. *The Journal of Chemical Physics* 65: 3006–3024.

Troup, G.J. and Hutton, D.R. 1964. Paramagnetic resonance of Fe^{3+} in kyanite. *British Journal of Applied Physics* 15: 1493–1499.

Tucker, E.B. 1966. Spin-lattice coupling of a Kramer's doublet: Co^{2+} in MgO. *Physical Review* 143: 264–274.

Uhlmann, H., Iametti, S., Vecchio, G., Bonomi, F., and Bernhardt, R. 1997. Pro108 is important for folding and stabilization of adrenal ferredoxin, but does not influence the functional properties of the protein. *European Journal of Biochemistry* 248: 897–902.

Van der Kraaij, A.M.M., Koster, J.F., and Hagen, W.R. 1989. Reappraisal of the e.p.r. signals in (post)-ischaemic cardiac tissue. *Biochemical Journal* 264: 687–694.

Van der Waals, J.H. and De Groot, M.S. 1959. Paramagnetic resonance in phosphorescent aromatic hydrocarbons. I: naphtalene. *Molecular Physics* 2: 333–340.

Van Niekerk, J.N. and Schoening, F.R.L. 1953. A new type of copper complex as found in the crystal structure of cupric acetate, $Cu_2(CH_3COO)_4 \times 2H_2O$. *Acta Crystallographica* 6: 227–232.

Venable, J.H. 1967. Electron paramagnetic resonance spectroscopy of protein single crystals: II. Computational methods. In *Magnetic Resonance in Biological Systems*, eds. A. Ehrenberg, B.G. Malmström and T. Vänngård Elmsford. New York: Pergamon Press, 373–381.

Verhagen, M.F.J.M., Kooter, I.M., Wolbert, R.B.G., and Hagen, W.R. 1993. On the iron-sulfur cluster of adenosine phosphosulfate reductase from *Desulfovibrio vulgaris* (Hildenborough). *European Journal of Biochemistry* 221: 831–837.

Von Waldkirch, Th., Müller, K.A., and Berlinger, W. 1972. Analysis of the Fe^{3+}-VO center in the tetragonal phase of $SrTiO_3$. *Physical Review* B 5: 4324–4334.

Wasserman, E., Snyder, L.C., and Yager, W.A. 1964. ESR of the triplet states of randomly oriented molecules. *The Journal of Chemical Physics* 41: 1763–1772.

Watanabe, H. 1966. *Operator Methods in Ligand Field Theory*. Englewood Cliffs, NJ: Prentice Hall.

Weil, J.A. and Bolton, J.R. 2007. *Electron Paramagnetic Resonance. Elementary Theory and Practical Applications,* 2nd edition. Wiley: Hoboken, NJ.

Weltner, W. Jr. 1983. *Magnetic Atoms and Molecules*. New York: Van Nostrand Reinhold. Reprinted Mineola, New York: Dover Publications, 1989.

Whisnant, C.C., Ferguson, S., and Chesnut, D.B. 1974. Hyperfine models for piperidine nitroxides. *Journal of Physical Chemistry* 78: 1410–1415.

Zener, C. 1951. Interaction between the d-shells in the transition metals. II. Ferromagnetic compounds of manganese with perovskite structure. *Physical Review* 82: 403–405.

Zeth, K., Offermann, S., Essen, L.-O., and Oesterhelt, D. 2004. Iron-oxo clusters biomineralizing on protein surfaces: structural analysis of *Halobacterium salinarum* DpsA in its low- and high-iron states. *Proceedings of the National Academy of Sciences of the USA* 101: 13780–13785.

Index

Printed in the United States
by Baker & Taylor Publisher Services